Praise for *Does Your Family Make You Smarter?*

"Another superb piece of work by the best mind in the business. The analysis of data is penetrating, the elaboration of its meaning highly illuminating, and the discourse on theories of intelligence is a feast for the mind."
Thomas J. Bouchard, Jr., Winner of the Dobzhansky Memorial Award for a Lifetime of Outstanding Scholarship in Behavior Genetics

"Another amazing analysis of IQ data by James Flynn! As author of the Stanford-Binet 5, I have admired Flynn's work for many years. I highly recommend his new book that shines new light on the life-course of intelligence."
Gale H. Roid, Author of *Stanford-Binet Intelligence Scales, Fifth Edition*

"James Flynn takes up one of the most important questions in the social sciences – What is left of human autonomy in the genomic age? – and lays out the optimistic case with full acknowledgment of the technical difficulties his argument must surmount. This is the way that we are going to make progress: by engaging an evolving state of knowledge with logic and data, transparently clear prose, and unfailing civility."
Charles Murray, Co-author of *The Bell Curve*

"Few intellectuals have grappled honestly with the problems surrounding the causes and effects of intelligence, and fewer still have done so with as much incisiveness and originality as James Flynn."
Steven Pinker, Psychology, Harvard University

"Professor Flynn has a remarkable ability to explain complex concepts in a way so rational and logical that it seems, after the event, we should be kicking ourselves for overlooking the obvious. His chapter on the Raven's Progressive Matrices is brilliant."
John Rust, Director of the Psychometrics Centre at Cambridge and co-author of *Raven's Progressive Matrices*

"James Flynn, as much as anyone, can take credit for ushering in the age of enlightenment in our understanding of the nature of human intelligence. In this latest chapter, we learn how our families can either advantage or disadvantage us, and how our choices can either foster or impede our intellectual performance."
Joshua Aronson, Applied Psychology, New York University

Praise for *What Is Intelligence?*

"Flynn is a cautious and careful writer. Unlike many others in IQ debates, he resists grand philosophizing . . . The evidence in support of Flynn's original observation is now so overwhelming that the Flynn effect has moved from theory to fact."
Malcolm Gladwell, *The New Yorker*

"This book is full of insightful ideas about our measuring rods and the ways in which they tap the thing that matters: the brain's relative capacity to use memory and learning to adapt to the world as we have made it."
The Times Higher Education Supplement

"Flynn paints a dynamic picture of what intelligence is and has produced an impressively multidimensional and often wise look at the elusive topic of human intelligence."
Publisher's Weekly

"A masterful book that will influence thinking about intelligence for many years to come."
Robert J. Sternberg, *PsycCRITIQUES*

"Mainstream IQ researchers, who are used to being demonized when they are not being ignored, admire Flynn, who is politically a man of the left, for his fairness, geniality, insight, and devotion to advancing knowledge."
Steve Sailer, *vdare.com*

"In *What Is Intelligence?* James R. Flynn ... suggests that we should not facilely equate IQ gains with intelligence gains. He says that it's necessary to 'dissect intelligence' into its component parts: 'solving mathematical problems, interpreting the great works of literature, finding on the spot solutions, assimilating the scientific worldview, critical

acumen and wisdom.' When this dissection is carried out, several paradoxes emerge, which Flynn in this engaging book attempts to reconcile."
Richard Restak, *American Scholar*

"In a brilliant interweaving of data and argument, Flynn calls into question fundamental assumptions about the nature of intelligence that have driven the field for the past century. There is something here for everyone to lose sleep over. His solution to the perplexing issues revolving around IQ gains over time will give the IQ Ayatollahs fits!"
S. J. Ceci, Cornell University

"This highly engaging, and very readable, book takes forward the Dickens/Flynn model of intelligence in the form of asking yet more provocative questions ... A most unusual book, one that holds the reader's attention and leaves behind concepts and ideas that force us to rethink all sorts of issues."
Sir Michael Rutter, King's College London

Praise for *Are We Getting Smarter?*

"... one of the most extraordinary science books I have ever read ... Flynn can confidently look forward to immortality. His name will forever be attached to one of the most contentious, baffling and, for me, exhilarating scientific discoveries of our age."
Bryan Appleyard, *The Sunday Times*

"Flynn has made this field his own ... This book's strengths are the authority of the author, the engaging writing style, the importance of the topics dealt with, and the up-to-date nature of the content."
Ian J. Deary, University of Edinburgh

"No one but James Flynn could have written this book. It contains his most recent ideas about the causes and implications of the massive rise in IQ test scores that has been termed the 'Flynn Effect', and is thus essential reading for anyone wishing to keep up to date with the latest thinking about the nature of IQ."
Nicholas Mackintosh, University of Cambridge

"The scholarship of this book is detailed and exhaustive. The originality of thinking is sprinkled throughout the beginning chapters, and reaches a peak in the final two. With his unique perspective, Flynn literally is 'opening new windows.'"
Jonathan Wai, Duke University

Does Your Family Make You Smarter?

Does your family make you smarter? James R. Flynn presents an exciting new method for estimating the effects of family on a range of cognitive abilities. Rather than using twin and adoption studies, he analyzes IQ tables that have been hidden in manuals over the last 65 years, and shows that family environment can confer a significant advantage or disadvantage to your level of intelligence. Wading into the nature vs. nurture debate, Flynn banishes the pessimistic notion that by the age of 17, people's cognitive abilities are solely determined by their genes. He argues that intelligence is also influenced by human autonomy – genetics and family notwithstanding, we all have the capacity to choose to enhance our cognitive performance. He concludes by reconciling this new understanding of individual differences with his earlier research on intergenerational trends (the "Flynn Effect") culminating in a general theory of intelligence.

JAMES R. FLYNN is Professor Emeritus at the University of Otago, New Zealand, and a recipient of the University's Gold Medal for Distinguished Career Research. He is renowned for the "Flynn Effect", the documentation of massive IQ gains from one generation to another. Professor Flynn is the author of 14 books including *Are We Getting Smarter?* (Cambridge University Press, 2012), *Where Have All the Liberals Gone?* (Cambridge University Press, 2008), and *What Is Intelligence?* (Cambridge University Press, 2007), which have caused many to rethink the prevailing theory of intelligence.

Does Your Family Make You Smarter?

Nature, Nurture, and Human Autonomy

JAMES R. FLYNN

CAMBRIDGE
UNIVERSITY PRESS

CAMBRIDGE
UNIVERSITY PRESS

University Printing House, Cambridge CB2 8BS, United Kingdom

Cambridge University Press is part of the University of Cambridge.

It furthers the University's mission by disseminating knowledge in the pursuit of education, learning and research at the highest international levels of excellence.

www.cambridge.org
Information on this title: www.cambridge.org/flynn

First published 2016

Printed in the United States of America by Sheridan Books, Inc.

A catalogue record for this publication is available from the British Library

Library of Congress Cataloguing in Publication data
Names: Flynn, James R. (James Robert), 1934- , author.
Title: Does your family make you smarter?: nature, nurture, and human autonomy / James R. Flynn.
Description: Cambridge ; New York : Cambridge University Press, 2016. | Includes bibliographical references and indexes.
Identifiers: LCCN 2015049112 | ISBN 9781316604465 (pbk. : alk. paper) | ISBN 9781107150058 (hbk. : alk. paper)
Subjects: | MESH: Intelligence--genetics | Family | Social Environment | Individuality | Intelligence Tests | Gene-Environment Interaction
Classification: LCC BF711 | NLM BF 431 | DDC 155.7-dc23 LC record available at http://lccn.loc.gov/2015049112

ISBN 978-1-107-15005-8 Hardback
ISBN 978-1-316-60446-5 Paperback

In memory of Nicholas Mackintosh (1935–2015)

Good friend–great critic

It matters not how strait the gate,
How charged with punishments the scroll,
I am the master of my fate,
I am the captain of my soul.
("Invictus", William Ernest Henley)

Contents

Contents

Appendix III: Raven's Progressive Matrices   203
(Appendices I–XIV online)

Figure

Tables

Boxes

Acknowledgments

My thanks to Richard Lynn, John Raven, and Gale Roid for providing data I used in my analysis of Raven's Progressive Matrices and the Stanford-Binet. My thanks to Richard Nisbett for answering questions about adoption studies. My thanks to Ian Deary, Bill Dickens, Roberto Colum, Branton Shearer, and Robert Sternberg for information about theories of intelligence. The new method used here was first outlined in *Intelligence and Human Progress: The Story of What Was Hidden in Our Genes* (Elsevier, 2013). The Syndics of Cambridge University Press were kind enough to allow me to paraphrase pages 203–219 from *What Is Intelligence?: Beyond the Flynn Effect* (2nd edition, 2009), and pages 284–286 from *Are We Getting Smarter? Rising IQ in the Twenty-First Century* (2012).

Part I

Human autonomy

1   Twins and autonomy

Who is correct? Those who believe that our family history and decisions affect our cognitive abilities, or those who cite twin studies to show that our intelligence is largely the product of our genes? This is really a debate about the limits of human autonomy.

Until kinship studies began to partition IQ differences between people in terms of what proportion was due to their genetic differences and what proportion was due to their environmental differences, most people thought of themselves as individuals whose personal life history, and personal decisions, made them unique. That did not mean that genes could be ignored. I knew very well I did not have the genes to become a Mozart or an Einstein but, just as Graham Greene said "England made me," I was convinced that my unique family history counted for what I was and that my personal decisions (to go to the University of Chicago rather than play it safe by going to the Catholic University of America) were significant.

It may be said, what could make you more uniquely yourself than your particular set of genes? But that is the problem: to lament your genes is to wish that you had been born a completely different human being. Genes cannot be personified in the image of yourself and your parents. You can love or hate your parents, be grateful or censorious about how they raised you, lament the injustice of a home in which poverty cheated you of advantages, exult in the freedom to choose your fate. The ownership of your unique past, present, and future is the essence of human autonomy.

The ownership of your genes is kismet and the categories of justice and freedom do not apply, unless you upbraid God because you were born at all.

I am not trying to create a straw man. Those who think our inherited genes overwhelm environment in the development of our cognitive abilities do not deny that family is important in many respects. They concede that parents affect whether their kids hate other races, get a criminal record, or learn to slap their own children, and indeed, they concede that family can give children a head start for cognitive abilities that counts in school and university.

The real question is whether family and personal choice have *long-term* significance for the development of *cognitive abilities* of the sort measured by IQ tests.

After all, at the age of 17 or 18, your cognitive abilities have a profound influence on your fate. By that age, some have failed to graduate from high school, and among those that have, most apply to universities whose quality does much to influence subsequent life history. These universities screen applicants for intelligence; that is, they look at your final set of grades at secondary school and how well you score on the SAT or Scholastic Aptitude Test, which is primarily a disguised IQ test. As an adult, your cognitive ability affects the peers you seek out as friends, your job performance, even whom you marry. Assume that the twin (or kinship) studies show that family effects on IQ have disappeared by ages 17–18 and therefore, genes dominate IQ. This is to say that whether you come from a bookish upper-class family or a typical working-class family is not relevant.

In *The Bell Curve*, Herrnstein and Murray (1994) note that liberals have tended to cast aspersions on the homes of ordinary people such as the working class. They have falsely assumed that those homes are so bankrupt in cognitive quality as to leave a permanent mark on the child's intellect. Note, however, the flip side of this conclusion: that working-class parents who spend so much time and money trying to duplicate the advantages of the typical middle-class home are prey to an illusion.

The message of the twins

A host of problems surrounds the family's influence on cognitive abilities. Studies of identical twins raised apart from birth are designed to separate genetic from environmental influences on IQ. If despite being raised in separate environments, the twins grew up to have identical IQs, we would know that their identical genes were all-powerful. If they grew up to have IQs no more alike than randomly selected individuals, we would know that environment was all-powerful. These studies are a fragment of a body of kinship studies that have the same purpose: comparing identical twins with fraternal twins (genes no more alike than brother and sister) when each twin pair is raised in the same home; comparing adopted children (whose genes would be unlike their adoptive parents) with their un-adopted brothers and sisters (who share genes with their parents).

This huge body of literature yields three factors that influence IQ differences between individuals: genes, family environment (sometimes called common environment), and "chance" environment (sometimes called uncommon environment), which is uncorrelated with both genes and family. Just as being raised in different homes has an independent influence on someone's cognitive abilities, at least in childhood, so do thousands of events that affect some people rather than others: being dropped on your head, being deserted by your spouse, unemployment, a death that sends you into depression, and so forth. These studies are virtually unanimous on three points.

First, *family has little effect on whatever cognitive abilities you have after the age of 17*. While family environment is potent early on, its effects fade away to a low level by age 17 and become insignificant by maturity. As you grow up, you move outside the family and go to school, become a member of a peer group (your close friends), find a job, and marry. You enter a current environment that swamps the lingering effects of family environment. Current environment is surprisingly self-contained: it influences one's current cognitive abilities with very little interference from past environments. Most of us assume that your early family environment leaves some sort of indelible mark on your intelligence throughout life. But the literature shows that this is simply not so.

Second, once the influence of family disappears, *the cognitive quality of your current environment tends to match your genetic quality*. This is often called a tendency toward "gene-environment co-relation." This means simply that if your genes are at the 90th percentile of the population for cognitive quality, your current environment tends to be at the 90th percentile for cognitive quality. It appears that high-IQ people seek out more enriched environments (for example, study more, join the book club, enter cognitively demanding occupations) and society tends to select high-IQ people out for more enriched environments (bright people befriend them, schools put them into an honors stream, law schools accept them). In other words, chance events aside, genes and current environment tend to match, so whatever genetic differences exist predict cognitive performance without any need to take current environment into account.

Third, as would be expected, *chance factors tend to be constant throughout life and account for about 20 percent of IQ differences*. In other words, the events of life history qualify the perfect match between genes and current environment. Being a bright person in a high-quality environment never inoculates you against good or bad luck. Even a merchant banker can find current environment debased by unemployment, a traffic accident, or the personal tragedy of a child gone astray. Eventually, I will make a case that the autonomous decisions of an individual fit into this category, and that they confer good and bad luck of a purposive kind.

The role of chance entails an important fact. The perfect match between genes and current environment holds for *groups* of people, not for every *individual*. Assume you have selected out a group of people at the 84th percentile of vocabulary performance. At age 30, the overall match between their performance and the richness of their vocabulary environment may be perfect but, thanks to chance, individual differences persist: some people will have an environment at the 84th percentile and others above or below that. Recently a friend in Auckland found that the leader of a gang had an IQ of 150: his gang certainly did not supply him with a vocabulary environment as rich as that. Presumably in his mind he had reached the pinnacle of status and will never aspire to be a university lecturer; and no profession is likely to invite him to apply for a job. He takes satisfaction in his moral superiority: he has robbed only a few people rather than the millions robbed by merchant bankers.

This does not mean that there are any lingering family effects. Whatever mismatch of genes and environment occurs at age 30 simply affects the match of *current* environment and IQ. If that mismatch was correlated with *family background*, it would show up as a persistent family influence – and it does not.

Luck and justice

What conclusions are we to draw from these findings? That is what the first half of this book is all about. *The Bell Curve* (Herrnstein and Murray, 1994) brought the results of the twin studies to a wide public and inspired a political dialogue about social justice. Most people believe that sheer bad luck should not cripple a person's life prospects. Some people do have bad luck in the genetic lottery – that is, they are born with genes that put them very low on the IQ scale. The individual is of course not to blame for this and humane ideals suggest that some kind of compensation is due. Everyone, conservative or liberal, believes that society should help them by giving them sustenance and special education.

Others with normal genes are born into families (and neighborhoods) that blight their lives. The mere fact that at maturity cognitive abilities generally match genetic promise does not mean we should do nothing to alleviate these conditions. To suffer as a child in an impoverished home is an evil in itself no matter what the eventual effects on intelligence: right and left differ only as to means – that is, how to strike a balance between the welfare state and the free market as a cure.

The mere fact that at maturity cognitive abilities generally match genetic promise does not mean neighborhood and peer group have no lasting effects on one's life. The girl who thanks to ignorance about contraception has a child at 16, the boy whose gang lands him a criminal record at the same age, they are marked for life despite their adult mental skills. Intelligence is not everything. Your childhood years can mean you start adulthood, not only with obvious strikes against you but also with attitudes (not aspiring to transcend gang leader) and emotions (race hate or racial resentment) and traits (escaping reality through drugs) that color your whole life. Upgrading schools is one method of alleviating these evils that right and left share. There is the usual difference about means: the balance between improving public schools and providing vouchers to offer more parents the choice of a private school.

However, recall that there is a special sense of injustice among those who believe that thanks to circumstances thrust upon them they have never lived life to the full. That thanks to family or neighborhood or school, they lacked the vocabulary or knowledge or understanding to go to a good university, and thereby make life-long friends or find a spouse among those who offer less pub talk and more serious talk, or qualify for a profession worthy of their talents. I speak with some feeling here. All but one of the males in the older generation of my family suffered to some degree from alcoholism and I suspect that (as they all left school between the ages of 11 and 14) this was due to a mismatch between their promise and the kind of education that might have enhanced their lives. Yet, I can testify that all of them were highly intelligent, perhaps as intelligent as their genes "intended"; but that was not enough.

In addition, we must not lose sight of the question of whether freedom or personal choice has consequences. If in adulthood, chance aside, genetic quality predicts cognitive quality, are individuals powerless to enhance their intelligence? Chance may put an individual above or below those grouped at his or her level of genetic performance, but chance is beyond our control. Luck is no substitute for human empowerment.

Finally, the twins pose an evidential problem. Dick Nisbett (2009) and others tend to believe that twin studies and adoption studies conflict. He cites data in which children from lower-class backgrounds who were adopted into upper-class homes profited greatly from the enhanced quality of their new families: these children gained almost 12 IQ points even though they were tested as late as age 14. Is it really plausible that family effects become nil by age 17 or soon after?

Beyond the twins

I will use a new method to supplement the findings of the twin studies for a whole range of cognitive abilities. In the light of these new findings, I will conclude the following. First, that *whatever families do to upgrade the cognitive quality of the home persists long enough to influence their children's fate at the crucial age of 17.* Second, that *whatever society does to upgrade the cognitive environment of children has the same consequences* (this of course is really a corollary of the first conclusion). Third, that genes and luck notwithstanding, *all of us, both in childhood and maturity, have the capacity to choose to significantly enhance our cognitive performance.*

To those who are ignorant of the twin literature, these conclusions will seem self-evident. And I should add that few of those who cite the twin literature would reject them outright. But they would stress that their significance is very limited, and caution me against encouraging naïve beliefs about the potency of family environment and choice. Well, here, the degree of significance is everything. I hope to shed some light on that: nothing that will restore belief in the "perfectibility" of man, but something that will show that genes allow environment and choice far more scope than those suffering from "post-twin pessimism" may be aware.

Toward a meta-theory of intelligence

Hitherto much of my work in psychology has been an attempt to analyze the significance of *generational trends* in cognitive abilities – that is, the so-called "Flynn Effect," or massive IQ gains from one generation to another, gains that totaled over 30 IQ points in the twentieth century. This is not to imply that "intelligence" is identical with IQ. But IQ gains are a measurable "symptom" of true cognitive gains and I have tried to describe just what those gains were.

The first half of this new book is an effort to clarify a different problem, that of *individual differences* in cognitive ability – that is, the significance of the fact that within a generation some people have superior abilities to others. Having achieved what I believe to be clarity in these two core areas of intelligence, I am emboldened to put my conclusions in the context of a theory of intelligence (one which will also find a place for intelligence in the area of brain physiology).

Therefore, the second half of this book opens with a chapter that uses my new method on a test (Raven's Progressive Matrices) that has a crucial role to play in the theory of intelligence. It also argues that intelligence needs something called a "meta-theory," concepts that offer scholars advice about how to investigate intelligence. And finally, it surveys a wide range of scientific theories of intelligence to see whether they are following the proper advice, and whether they are compatible with one another.

To aid the reader, every chapter will be prefaced with questions to be addressed and end with those questions answered. The most difficult material concerns my new method for measuring family effects on various cognitive abilities. I will try to put it as plainly as possible and consign the detailed calculations that it entails to a series of fourteen appendices. The whole package is online at www.cambridge.org/flynn. The three that I consider most essential are published here. To aid researchers, at the end of the text (after Chapter 11), I have added a list of nations worth exploring to see if they have the proper test manuals to apply the new method to nations other than the US.

2   Justice and freedom

Questions

(1) When 17-year-olds take the SAT, do some homes enhance cognitive performance at this age more than others?
(2) After family effects are gone, can adults enhance their cognitive abilities?

At the age of 17, cognitive performance does much to determine the fate of American youth. That is the age at which they take the SAT and sort themselves out among various universities. I am going to ask you (for now) to accept two promissory notes: that we know how much family affects vocabulary at various levels of achievement; and that we have a rough estimate of the percentile gap between levels of achievement and the cognitive quality of the family typical at that level. For example, those at the 98th percentile of vocabulary come on average from homes just below the 70th percentile of cognitive quality. Assuming we have this knowledge, let us look at the consequences.

Vocabulary and family quality

When students sit the SAT, universities take scores on the SAT for reading (SAT-R) as the best measure of the viability of their students. Vocabulary is highly predictive of those scores. I will average Vocabulary results from all six of the data sets the leading IQ tests give us: Stanford-Binet tests from 1985 and 2001; Wechsler tests from 1950–55, 1975, 1992, and 2004–05 (these dates average the years when the Wechsler Intelligence Scale for Children and the Wechsler Adult Intelligence Scale, the WISC and the WAIS, were actually normed). By combining their results, I hope to eliminate the vagaries of any particular set.

Table 1 shows that family has different effects at various levels of performance. For example: students whose vocabulary puts them at +2 SD above average, which is the 97.73 percentile (they are better than almost 98 percent of 17-year-olds), suffer from a typical disadvantage thanks to their families of about 1 IQ point; those at +1 SD (or at the 84th percentile – better than 84 percent of their peers) have a typical disadvantage of 3 IQ points; typical students at −1 SD (the 16th percentile) are advantaged by just under 3 points: while those way down at the −2 SD (the 2.27 percentile) are advantaged by over 7 IQ points.

I know the way the pluses and minuses are used can be confusing. Imagine that society is keeping the books: a (+) is like a fine that society collects to punish you for having a family whose cognitive value is worse than your level of performance; a (−) is a

Table 1 Vocabulary: typical family disadvantages/advantages at ages 17–18

	Wechsler 1950–55	Wechsler 1975	Wechsler 1992	Wechsler 2004–05	SB 1985	SB 2001	Average
+2 SD	+2.90	−2.14	−4.21	+0.75	+6.37	+2.14	**+0.97**
+1 SD	+6.41	+1.43	+1.37	+5.25	+3.56	0.00	**+3.00**
−1 SD	−2.31	−1.25	−3.83	−4.18	−2.91	−2.14	**−2.77**
−2 SD	−8.38	−10.26	−8.39	−5.25	−6.85	−4.29	**−7.24**

+ denotes a disadvantage; − denotes an advantage; read on as to why.

rebate society sends you to reward you for having a family whose cognitive value is better than your level of performance.

This makes perfect sense. It would be incredible if those at the 98th percentile of performance came from homes that averaged at the 98th percentile on a cognitive quality of environment scale – that is, homes almost exclusively in the top 5 percent. As we all know, many high-achieving students come from non-elite homes, perhaps few from welfare homes but many from middle-class and working-class homes – not all of them are the children of academics. I will eventually argue that, on average, they come from homes just below the 70th percentile in terms of cognitive quality. Clearly, that is far below the average level of their genetic promise (which must be at least slightly above the 98th percentile: otherwise how could they score so high?). Therefore, on average, they suffer from a family handicap. That it is as little as 1 IQ point shows how much family effects have faded by age 17 *at this level.*

The level is important. Someone of this high genetic promise will tend to match a current environment that is almost equally high after they go to school. They will be highly articulate and attract the teacher's attention, learn to read quickly and join the library club, make articulate friends who reinforce their vocabulary, join an honors stream, and by the age of 17, the initial effects of their family environment will be swamped by a new current environment of very high quality.

It is a different story at other levels of vocabulary performance. At the 84th percentile, which is still quite high, they will still have an average home environment below that level: many will come from homes in the bottom half of cognitive quality. After they attend school, their good performance will still tend to replace family effects with a vocabulary environment that comes closer to their genetic promise. But the trend is less pronounced. Family effects linger on and at 17, levy a penalty of 3 IQ points. This means that if their home environment matched their genetic promise, the typical student at this level would rise from a "Vocabulary IQ" of 115 to 118. I say 115 because the average score is by definition 100 and each SD is worth 15 points. Thus at +1 SD above the median they are performing at a 115-IQ level.

Below average performers will be the mirror image of high performers. Those performing at the 2nd percentile will on average come from homes that are well above that for cognitive quality (some of them will be from elite homes). And here is something encouraging: after they go to school, special attention keeps them from sinking down to a cognitive environment close to their low level of genetic promise. No doubt, they cannot access the rich environment school offers high achievers. However, they get a quality of school environment well above the 2nd percentile level, a current environment that may eat away at the advantage their typical homes confer, but does not entirely obliterate it. After all, they are still living at home and interacting with parents and siblings whose vocabulary is better than their own. At 17, they retain a family advantage of fully 7 IQ points. It would be interesting to know what the situation was like before these students were mainstreamed into ordinary classrooms, rather than being segregated as in the past. Those performing at −1 SD (16th percentile) for Vocabulary typically have a family advantage of 2.73 IQ points at age 17. They must have a school experience very like those +1 SD above the median who retained a family disadvantage of 3.00 points.

Table 2 requires (for now) another act of faith. It assumes I can justify at least rough estimates of the average level of family quality that are appropriate at various levels of Vocabulary performance. The values in bold are my estimates: those at +2 SD (remember that is equivalent to an IQ of 130) on average come from homes at the 69th percentile of cognitive quality: those at +1 SD (IQ 115) from homes at the 61st percentile, those at −1 SD (IQ 85) from homes at the 39th percentile, and those at −2 SD (IQ 70) from homes at the 31st percentile.

Table 2 Ages 17–18: percentile quality of home and score at that percentile for four performance levels (Vocabulary)

+2 SD	98th: 130.97	**69th: 130.00**	34th: 129.03	—	1st: 128.06
+1 SD	84th: 118.00	**61st: 115.00**	33rd: 112.00	12th: 109.00	1st: 106.00
−1 SD	89th: 90.54	68th: 87.77	**39th: 85.00**	16th: 82.27	4th: 79.46
−2 SD	84th: 77.25	(66th: 73.50)	**31st: 70.00**	(17th: 66.38)	2nd: 62.75

Calculations:

(1) +2 SD score = 130 with a handicap of 0.97 points for being at the 69th percentile (bottom 30% gone = mean at +0.4967 SD gives 69th percentile) rather than the 97th. Now 2.00 SD − 0.4967 = 1.5043 SD above and *that* costs them 0.97 points. Taking 1.5043 from 0.4967 = 1.0076 below the median or the 34th percentile. Adding 1.5043 to that = 1.0076 + 1.5043 = 2.5119 below or the first percentile.

(2) +1 SD score = 115 with a handicap of 3 points for being at the 61st percentile (bottom 15% gone = +0.2743 = 61) rather than the 84th. Now 1.00 SD − 0.2743 = 0.7257 SD above and *that* costs them 3 points. Taking 0.7257 from 0.2743 = 0.4514 below the median or 33rd percentile. Adding 0.7257 to that = 0.4514 + 0.7257 = 1.1771 below or the 12th percentile. Adding 0.7257 = 1.1771 = 2.4967 below or the 1st percentile.

(3) −1 SD score = 85 with a boon of 2.77 points for being at the 39th percentile (top 15% gone = −0.2743 = 50 − 11 = 39) rather than the 16th. −0.2743 − 1.00 SD = +0.7557 and *that* gave them extra 2.77 points. −0.2743 + 0.7557 = 0.4814 above the median or the 68th percentile. 0.4814 + 0.7557 = 1.2371 above or the 89th percentile. One SD below the median + 0.7557 = 1.7557 below or the 4th percentile.

(4) −2 SD score = 70 with a boon of 7.25 point for being at the 31st percentile (top 30% gone = −0.4967 = 31) rather than the 2nd percentile. Now −2.00 + 0.4967 = 1.5043 and *that* gave them 7.25 points. −0.4967 + 1.5044 = 1.0077 above the mean or the 84th percentile for an extra 7.25 points. To lose 7.25 points they have to +1.5044 down from the typical, which was worth +1.5043. So their home environment must essentially match their performance: the 2.27th percentile.

(5) The values in brackets are interpolations but are reasonably accurate.

Table 1 gave estimates only for those who suffer or benefit from the typical distance between their level of performance and the cognitive quality of their homes, the high performers suffering disadvantages, the low performers enjoying advantages. Table 2 takes into account that students at all levels will actually come from a range of homes in terms of cognitive quality. For those who score 130 for Vocabulary, this makes little difference by the age of 17, because remaining family effects are so small.

When they take the SAT-R, their homes could be anywhere from the 98th percentile of cognitive quality all the way down to the 34th percentile, and the difference to their vocabulary would be less than the equivalent of 2 IQ points. But they are only the top 5 percent of applicants. Those at other levels are not so fortunate.

As Table 2 shows, if the typical person who scores 115 (+1 SD) happened to come from a home equivalent to their genetic promise, they would have scored 118. But if they had the bad luck to come from a home at the 12th percentile of cognitive quality, they would have scored only 109, or 9 points less.

Almost everyone whose vocabulary is at 115 (the 84th percentile) would think of themselves as college material. As Flynn (2013) points out, every 3 IQ points of Vocabulary IQ is equivalent to about 22 SAT-R points (IQ SD = 15; SAT SD = 110). A 9-point IQ deficit equals 66 SAT-R points. Take those who would have a Vocabulary of 115 if they came from a home typical of those at that level (61st percentile). If they came from a privileged home environment (84th percentile: perhaps an academic home), they would have a Vocabulary IQ (118) which translates into an SAT-R score of about 566; someone from a bad home environment (12th percentile) would have a Vocabulary (109) which translates into an SAT-R score of about 500. Those 66 points are terribly important. Many know the despair caused by the receipt of SAT scores that shut the door on a student to the university of their dreams.

Vocabulary and the universities

Since universities believe SAT-R scores determine which students will be viable, they make public the SAT-R score that isolates the bottom 25 percent of their intake. Table 3 shows the full impact of family environment on university prospects.

Table 3 Family environment and college viability (age 17)

Level	At +1 SD of Vocabulary performance			
Family environment (percentile)	84th	**61st**	33rd	12th
Vocabulary IQ	118.00	**115.00**	112.00	109.00
SAT-R	566	**544**	522	500
Level	**At −1 SD of Vocabulary performance**			
Family environment (percentile)	68th	**39th**	16th	4th
Vocabulary IQ	87.77	**85.00**	82.27	79.46
SAT-R	344	**324**	304	284

	25th percentile SAT-R at selected universities
Brigham Young (Utah)	570
Pittsburg (Pennsylvania)	570
UCLA (California)	570
U. Florida	570
	SCORE OF 566
Baylor (Texas)	560
Beloit (Wisconsin)	560
U. Georgia	560
Clemson (South Carolina)	550
Florida State	550
U. Connecticut	550
U. Denver	550
	SCORE OF 544
Ohio State	540
U. California San Diego	540
U. Delaware	540
U. Maryland (Baltimore)	540
U. Minnesota	540
U. Texas (Austin)	540
U. Vermont	540
Virginia Tech	540
	SCORE OF 522

SCORE OF 500: viable at typical US university	
	SCORE OF 344
25th percentile SAT-Reading at selected universities	
Dakota Wesleyan	340
Oklahoma Panhandle	340
Upper Iowa	340
Presentation College (SD)	330
	SCORE OF 324
Tougaloo (Mississippi)	320
	SCORE OF 304
	SCORE OF 284
Faulkner (Alabama)	281

Adapted from Table 9 of Flynn (2013), with permission from Elsevier Publishers.

The first half of the table focuses on those who would get an SAT-R score of 544, if they came from a home typical of those who score at that level. In fact at age 17, they would actually range from 566 down to 500 according to the cognitive quality of their homes. The second half of the table puts these SAT-R scores in the context of viability at selected US universities. Those from the 84th percentile of home quality get 566, and are near viable at universities as distinguished as UCLA. They are fully viable at very good universities such as Baylor, Beloit, and the University of Connecticut. Those from the 61st percentile get 544 and must shoot a little lower at universities like the University of California at San Diego, the University of Minnesota, and the University of Texas at Austin. Those from the 33rd percentile get 522 and will not be viable at a top university. Those from the 12th percentile get 500 and must settle for the average university.

In America, contrary to most other advanced nations, there are universities that cater even for those at the 16th percentile of Vocabulary performance. Therefore, Table 3 isolates those who would typically have an SAT-R score of 322 and shows that their scores would actually range from 344 down to 284 according to the cognitive quality of their homes. Those from the 68th percentile of home quality are viable at universities whose standards tend to match this group, such as Dakota Wesleyan and Oklahoma Panhandle. Those from the 39th percentile had better live in Mississippi (which has Tougaloo University) or Alabama (Faulkner University). Those from the 16th or 4th percentiles miss out even on the least demanding university I could find.

I do not mean to emphasize university entrance above all else (it is just more quantifiable). As Table 3 shows, at −1 SD, a range of environments may determine whether you find school challenging or almost hopeless: whether your Vocabulary "IQ" is 91 rather than 82. At −2 SD, family environment can determine whether you can avoid or must accept the crucial label of suffering from mental retardation: whether your Vocabulary "IQ" is 77 rather than 63. Those aware of the external validity of a vocabulary functioning at plus or minus 9 points (or even 14 points) will think of many examples that either advantage or disadvantage.

Rough estimates

When we justify my estimates of the typical gap between levels of Vocabulary performance and the quality of the homes from which these levels come, they will prove to be rough. However, note that any revision of my estimates is a double-edged sword. I suspect that most will think I have put the performance/home-quality gaps too high. If that is so, the real-world consequences for those below the median will be less serious, the consequences for those above the median will be more serious, and the consequences for the average person will be unaffected.

Remember that the sizes of the *typical* score disadvantages of those above the median are set by the analysis of the age tables. Whatever the discrepancy between the quality of performance and home may be, those who perform at the 84th percentile are handicapped by exactly that score disadvantage (3 IQ points) – if they are from the average home for those at that level. The guesses at the size of the unusual performance-home discrepancies are used only to calculate the consequences for atypical individuals whose home may be better or worse.

Let us assume that those who perform at the 84th percentile (115) typically come from homes at the 75th percentile of cognitive quality rather than the 61st (my guess). That means they lose 3 points despite being more privileged than I think them to be. And this would raise the percentiles of the homes that handicap them all the way down the line. Rather than losing 3 points from a home at the 61st percentile and 9 points from a home at the 12th percentile, they would lose the same points by coming from homes at higher percentiles. They would lose their 3 points at the 75th percentile (by definition) and their 9 points at the 23rd percentile. On the other hand, if my guesses of the typical performance/home gaps are too high, those below the median would profit. Take those whose performance puts them at 70 if they are from homes typical for that level. Once again the typical are unaltered. But the atypical lose their 7 points not by being from a home at the 2nd percentile (my guess), but by being from a home down on the 1st percentile.

Consequences after university

Someone may take solace in the fact that by mid-adulthood (assuming luck does not single them out), his or her current environment for vocabulary will closely match his or her genetic promise. But people cannot relive their childhood and university years. Whether or not a substandard vocabulary made school a continual struggle leaves a mark. Being put in special education can be a mixed blessing. What university they attended may determine their occupation, whether they have influential friends, their choice of spouse, things not easily altered (except the spouse, perhaps).

Most of us love our families. But there is no doubt that the family lottery levies disadvantages and advantages that, from the individual's point of view, are unjust.

Autonomy after university

University does not (always) kill a person's desire to upgrade their cognitive abilities. However, if by adulthood genes and current environment are perfectly matched, how is this possible? Your genes seem to dictate the environment you have got, your present cognitive performance is the result, so what is the point of trying? There is one thing about the twin studies that has been overlooked: they show that about 20 percent of IQ variance is due to chance environment throughout life. Fortunately this fact is not in dispute (Haworth et al., 2010). It is its significance that needs to be explored. The key is that chance environment is really composed of two things: what accidentally happens to you, and what you make happen to you.

The 20 percent does cover events beyond our control: accidents, illnesses, sudden unemployment or family breakdown, a lucky appointment to a job that challenges us, events that have little to do with our genetic promise. However, it also evidences that *human autonomy* has important effects: we can actually choose to alter our cognitive environment so that it either transcends or falls short of our place on the genetic hierarchy. It is possible that this could go on without engendering any IQ variance explained whatsoever. Every time bad luck or good luck alters our cognitive environment so that the match between our genes and our current environment is broken, we might make a choice that exactly redresses the balance. That is unlikely to happen very often. It is good that we have 20 percent of variance available as evidence that this kind of equivalence is not always attained. It shows that powerful effects are at work, and at least some of these effects can be the product of individual choice.

The immediate impact of current environment

The potency of an individual's choice of a better environment is dependent on direct evidence as to just how much current environment affects cognitive performance. If the new environment's impact had to be averaged with the influence of a series of past environments, enhancement of cognitive performance would be delayed. As Bill Dickens has pointed out (cited in Flynn, 2007, pp. 97 and 99), to get direct evidence you would need to: collect a large sample with over-representation of identical twins; accumulate data on their occupations, hobbies, and friends; and test their IQs yearly with due attention to subtests.

This has never been done, but sometimes the world presents you with a good experimental design. I refer to Adam et al. (2007). They compared performance on a test of episodic memory between two age groups – males aged 50 to 54 and 60 to 64 respectively – they ranked twelve nations in terms of persistence of employment into old age. This cross-national comparison eliminated the obvious confounding variables. Within a nation, those who stayed in work would select out those who felt most intellectually alert or suffered least from decline with age. But between nations, we find contrasts in retirement age that have little to do with the intellectual or physical vigor of Frenchmen versus Swedes. When the percentage of those in work dropped by 90 percent (France), there was a 15 percent memory decline; when they dropped by 15 percent (Sweden), the decline was only 7 percent. Clearly the current work environment over those ten years had a dramatic effect that made past environments irrelevant. Episodic memory is not intelligence. We await studies that test for a wider range of cognitive abilities.

Partitioning the 20 percent

We do not know how to partition the 20 percent of variance between chance and choice, but it is clear that the autonomy portion is substantial. Take two Americans, aged 50, whose place in the genetic hierarchy exactly matches their place in a hierarchy of current environments ranked according to cognitive quality. One is disabled by an accident and gives up law to watch TV (particularly films that make us feel we are becoming dumber by the minute, such as *Dead Poets Society*); the other continues to practice. Now take another two men much the same. One voluntarily retires from law early to play golf, the other chooses to continue to practice and decides to write a book on jurisprudence. Clearly choice can break the match between genes and current environment just as decisively as chance. To illustrate how much potency choice may have, I will assume that it accounts for 10 percent of cognitive variance (half of the 20 percent available).

Table 4 estimates the potency of autonomous choice in terms of upgrading or debasing the cognitive quality of one's

Table 4 Autonomous individuals: effects of upgrading/degrading their current cognitive environment

	Cognitive quality of environment					
	99th	98th	84th	50th	16th	2nd
+2 SD	131.58	**130.00**	125.26	120.52	—	—
+1 SD	—	119.74	**115.00**	110.26	105.52	—
Median	—	—	104.74	**100.00**	95.26	—
−1 SD	—	—	—	89.74	**85.00**	80.26
−2 SD	—	—	—	79.48	74.74	**70.00**

Calculation: (1) If 10 percent of IQ variance is explained by current environment, the correlation between IQ and environment is 0.316 (the square root of the variance explained). (2) All the quality levels of current environment (save one) are set at so many SDs above or below the median: +2, +1, −1, −2. So (with SD = 15), take 15 points times 0.316 and you have an IQ shift of 4.74 points. Add or subtract that amount accordingly. (3) I have assumed that those whose genic quality puts them at the 97.73 percentile (98th in the table) would not be able to upgrade their current environment above the 99th percentile. This puts them 2.33 SD above the mean and since their typical place is 2 SD above, they would gain only 1.58 points (0.33 SD = 5 IQ points × 0.316 = 1.58 points).

current environment. I have made the effects of various environments on cognitive performance uniform at all performance levels from +2 SD down to −2 SD. I have also used the fact that if 10 percent of IQ variance is explained by a factor, the correlation coefficient between IQ and that factor will be 0.316. The square root of 0.10 (10 percent) gives 0.316 as the correlation. Take that on faith (the great Gauss vouched for it). As calculations at the bottom of the table show, every shift of 1 SD of environment (say from the 84th percentile to the 98th percentile) is worth 4.74 IQ points.

The table uses bold to designate what IQ you would have with a perfect match between genes and current environment. As you can see, those whose genetic "potential" is an IQ of 130 given a perfect match with current environment can easily drop to 125.26 if they retire (go from the 97.73th down to the 84th percentile of environmental quality) or to 120.52 (if they get very lazy and go down to the 50th percentile). Someone whose genetic "potential" is an IQ of 130, given a perfect match with current environment, may be so unlucky as to be almost 5 or 10 points below that because of a dead-end job and the peers that job provides. On the optimistic assumption that universities help you upgrade the cognitive quality of current environment, if someone becomes a mature student, he or she can rise from an IQ of 120.52 to 131.58 – that is, go from the 91.45 percentile of cognitive performance up to the 98.24 percentile. This means that they have leapfrogged over 82 percent of those who were once above them.

An upgrade of current environment would benefit people all the way down the IQ scale. Someone who could qualify for a training program for an elite job with an IQ of 115 may score barely above average (105) if seriously crippled by current environment. A person whose typical match would give an IQ of 70 could go up to 75 and be more likely to be employed. This assumes that those who enhance their current environment sustain their efforts, although the job itself may be enough. Note that no adult upgrading or downgrading current environment, whether above or below what quality of environment is correlated with genes, does anything to perpetuate the effects of family environment. The very reason it qualifies for the "chance" component of IQ variance is that it is not correlated with either genes or persistent family effects.

Within generations and between generations

I should add that the limited role (only 20 percent) accorded to chance environment (environment uncorrelated with genes) by partitioning variance *at any given time* does not circumscribe the potency of upgrading IQ or cognitive performance of generations *over time*. At any given time, you are correlating the genetic hierarchy with the hierarchy of environments available at that time, and the quality of the whole spectrum may be low. Therefore, if the whole hierarchy of environments gets a qualitative boost over time that will in itself greatly improve cognitive performance from one generation to another (Flynn, 2007).

Thus, we have the "Flynn Effect", the massive IQ gains of 30 or more points that have occurred thanks to social evolution over the last century. The fact that partitioning of variance within a generation gives environment uncorrelated with genes a minor role at maturity in no way compromises the powerful effects of environment. Its explosive potency is masked when it merely reinforces genetic differences but becomes evident when it operates free of genetic upgrading, as it does between generations.

Some of the subtests (Vocabulary, Information, Comprehension) that show children gaining the least from one generation to another are also the subtests that show the most persistent family effects within a generation. These are the very subtests most important for academic achievement and, therefore, their persistent family effects do most to handicap (or benefit) 17-year-olds. The advantage of breaking global IQ trends down into subtest trends is vindicated herein. Adults, by the way, have gained greatly on vocabulary over time for reasons I will make clear eventually.

Justice and autonomy

My previous books have celebrated the fact that the twins in no way circumscribe the march of humankind to a world that is more cognitively rich and morally improved (Flynn, 2013). However, this book focuses on the individual human being who lives within a generation and wants to know the effect of the environments available at that time on his or her cognitive abilities. Most have conceded that human progress is possible but many suffer from what I call "post-twin pessimism" about the plight of the individual.

More important to me, it sheds light on social questions. For example, it legitimates the individual's sense of suffering penalties inflicted by a home environment over which he or she had no control. This violates "justice as fairness," which all concede to be the essence of injustice. As I have said, we have excellent reasons for addressing environmental inequities, cognitive ability aside; but the fact that family effects still advantage or disadvantage cognitive performance at the crucial age of 17 adds its own rationale.

More important still, my analysis gives human autonomy a potent role. Here we must distinguish between external and internal environment. You can join the book club but it is more important to fall in love with reading; you can fill your mind with trash or ponder over chess problems or any other problem that provokes wonder (why politicians are so corruptible). When you upgrade the cognitive content of your mental life, you create a sort of portable cognitive gymnasium that exercises the mind. It is an environmental enhancement you can always carry with you despite adverse circumstances. Stephen Hawking dwells on physics despite a physical disability that would cause most of us to simply give up. My old professor Leo Strauss never seemed to think about anything but political philosophy from the moment he awoke. This is not a prescription for sanity. But a young man who goes into the army still reading and playing chess degenerates much less than someone who has no resistance to a mental climate of unquestioning obedience. He is like an athlete determined to run and maintain his physical fitness even when his competitive days are over.

As for using autonomy productively, children at school who try harder than most can upgrade their cognitive environment and reap important benefits. They can find within the 10 percent of autonomy variance (which holds at any age) the same benefits that adults do. They can enhance their Vocabulary "IQ" and read better than most and learn more than most. Whether they keep a cognitive ability edge later in life is up to them.

How wonderful it is that adults enjoy autonomy throughout their lives! University students come to me and say, "I know I am not as quick as the very best but I want to improve my mind and solve problems that captivate me; is that possible?" To this the answer is "yes." The very reason that mature students come back to university is often to escape a barren current environment. They say, "I did not do well at school; will I be unable to handle your introductory course in moral philosophy?" To this the answer is that you may do very well indeed: some of my best students are mature students because they work out of genuine interest. Note my assumption: that current environment is the key and they need not worry too much about the past environments that have handicapped them since school.

The twins and optimism

This highlights an optimistic finding of the twin studies. It is liberating that family effects are virtually nil as a cognitive influence in adulthood. Who would want to carry such a handicap throughout life? The fact that family effects fade frees us to get maximum benefit from our autonomy.

Needless to say, I do not admire only those who worship self-improvement. Women who go from child minding back to law or journalism probably value the fact that work gives their lives a transcendent purpose, rather than being concerned overmuch with upgrading their Vocabulary performance. Some may believe that they or their families would benefit from having the extra money. Those who retire at 65 may worry little about whether they lose a half standard deviation of IQ; they may take pleasure in more time to see their grandchildren. It is easy for scholars to become obsessed with their subject.

Answers

(1) Whether the family affects Vocabulary performance at the age of 17 is heavily dependent on genetic quality. For the top 5 percent, it has little impact. For those at 1 SD above or 1 SD below the median, it is typically worth about 3 IQ points. For those at 2 SD below, it is powerful and confers a typical advantage of 7 IQ points.

When students take the SAT, the "accident" of family environment has a profound influence on what university they attend; and for those whose skills are more modest, the environment has an influence on what they learn at school and whether they will be classified as mentally retarded.

(2) All of us at every age can aspire to upgrade our cognitive skills. Fully 20 percent of environment is not correlated with either genes or current environment. This leaves personal autonomy with a powerful role, in that you can choose to upgrade your current environment with important consequences for your cognitive abilities.

3   The great debate

Questions

(1) What evidence proves that family effects are still significant at ages 17–18?

(2) How can this data be converted into the number of IQ points that family environment confers as an advantage or disadvantage?

I will settle the great debate about whether family effects are still significant at the age of 17. We now know the practical significance of this question. It may seem trivial whether 3 IQ points hamper or help the typical 17-year-old thanks to his or her family environment, but this determines whether or not young people qualify for the universities to which they aspire and whether justice enters into the equation. Thus far the combination of twin studies and adoption studies have not been decisive. Fortunately, a new source of data decides the question. At least some cognitive abilities really are influenced by family environment at that crucial age.

A pause for reflection

To prepare the mind to appreciate the new method, the following is a useful exercise:

(1) Assume that individual differences on IQ tests and subtests are determined by two factors that are mutually

exclusive: differences in genes and differences in systemic environment like family. For the moment we will set aside differences in chance environment (life's boons and ills and the effects of human autonomy) as irrelevant.

(2) We want to compare childhood ages with the age of 50. Assume that by that age, genes and current environment have become perfectly correlated and family background has no independent influence. While at childhood, it does have an impact independent of genes.

(3) Surely it would make sense to compare performance at age 50 (when the two factors are "pulling together") with age 10 (when they are at odds) to see if we can detect a difference between the two that would allow us to measure the *degree* to which family environment has an independent impact.

(4) To follow through, it remains to isolate the proper data, to hypothesize how family might affect that data, to look for its footprint, and to find a way of measuring the size of that footprint.

Promising data

Vocabulary is the most important cognitive skill influenced by family. Therefore, I will begin with data from the Stanford-Binet (2001) Vocabulary subtest. It has the advantage that it gives tables from age 2 to age 90, and furnishes raw scores for every level of ability from 2.67 SD below the median (the bottom 0.4 percent) to 2.67 SD above the median (the top 0.4 percent). Raw scores are awarded for the number of items you get correct. I do not include scores at 3 SD above or below, for reasons given in Appendix II.

When family differences are independent

Take all those at age 6 whose Vocabulary level is at the 99th percentile or the top 1 percent. How likely is it that the cognitive quality of their homes averages at the top 1 percent? Surely many of them will come from homes at the 95th percentile, or even below the 70th percentile. If this is so, their performance will on average be a result of higher-quality genes (than the top 1 percent) being dragged down by lower-quality homes. And the same would be true at the bottom: surely the bottom 1 percent in Vocabulary performance would tend to come from homes somewhat better than that, which means they are being dragged up by homes whose average cognitive quality is better than their genes. In both case we would expect genes and environment to fall short of a perfect match.

Here, note something of supreme importance: typical performers at the median or 50th percentile would on average come from as many homes above that level as below. Even if family environment were potent at age 6, its effects would be muted at the median and weigh in only as we looked at those at higher or lower percentiles.

The family footprint

The age tables in the Stanford-Binet manual prove that the above speculations are correct. But to understand their crucial significance, you must choose an age at which you believe there is a perfect match between genes and family environment. I always choose the age at which raw score performance peaks, in this case ages 50–59. Later I will produce evidence that the influence of current environment on performance does cease at (or before) that age. But if not, making this assumption is a bias unfriendly to the size of environmental effects at earlier ages.

Then you calculate at younger ages the magnitude by which (raw score) performance falls short of the peak performance year. If our speculations are correct, as you get to the top percentiles, the gap between say age 6 and ages 50–59, will be *larger* than average. After all, 6-year-olds at the top 1 per cent (who are being handicapped by their homes) are being compared to adults that suffer from no handicap (family environment having faded away). And as you get to the bottom percentiles, 6-year-olds will be *closer* to the 50–59-year-olds at that level. The former are benefitting from a quality of family environment above their genes, the latter get no benefit from a family environment that has faded away. But at the median, the 6-year-olds get no handicap or benefit even if family environment is potent: their average genes are still matched with an average family environment.

Table 5 shows the raw score gaps that separate ages 2–18 from the target age (at which family environment is assumed to have faded away in favor of a perfect match between genes and environment). The progression from smaller gaps at low levels of achievement to greater gaps at high levels of achievement is extraordinary. Even at age 18, although the lowest achievers are only 0.5 points worse than the lowest achievers of the target age, the highest achievers are fully 4.5 points worse than their coun-terparts. Also note the values that cluster around the median (in bold). For ages 6–18 there is hardly any difference at all, evi-dencing the virtual absence of a mismatch between genes and environment. Even if family environment is potent, those at the median will come from as many homes above average as homes below average. The gap at the median represents merely improved performance as children age, with no extra bonus or handicap because of the effects of superior or inferior family environment. At age 16, the values around the median are actually random.

Any alternatives to family?

I am happy to have others suggest alternative hypotheses for the extraordinary character of the data. I will label it the "pattern of pro-gressive gaps." Surely it calls out for some explanation from anyone who rejects my own: the smaller and larger raw score gaps below and above the median, and the absence of any movement near the median, simply must be accounted for. My case rests on assumptions

Table 5 Pattern of progressive gaps

Manual levels	SD levels	Raw score gaps (with target age) by age and level of achievement								
		18	16	14	12	10	8	6	4	2
2	−2.67	0.5	1.50	3.00	4.50	6.00	9.50	15.25	20.25	25.0
3	−2.33	1.0	2.00	3.50	5.25	6.75	10.50	16.00	21.25	26.5
4	−2.00	1.0	2.00	3.50	5.50	7.00	10.75	16.25	21.75	27.5
5	−1.67	1.5	2.50	4.00	6.50	7.75	11.75	17.25	22.75	29.0
6	−1.33	2.0	3.00	4.75	6.50	8.50	12.50	18.25	24.00	30.5
7	−1.00	2.0	3.00	5.00	6.75	8.75	12.75	18.75	24.75	31.5
8	−0.67	2.5	3.50	5.75	7.75	9.75	13.75	19.75	25.75	33.5
9	**−0.33**	**3.0**	**4.00**	**6.50**	**8.50**	**10.50**	**14.75**	**20.50**	**26.75**	**35.5**
10	**Median**	**3.0**	**3.75**	**6.50**	**8.50**	**10.50**	**15.00**	**21.00**	**27.25**	**36.5**
11	**+0.33**	**3.0**	**3.50**	**6.50**	**8.50**	**10.75**	**15.25**	**21.50**	**27.75**	**37.5**
12	+0.67	3.0	4.25	7.00	9.00	11.75	16.25	22.50	29.00	39.0
13	+1.00	3.0	5.00	7.50	9.50	12.50	17.00	23.50	30.25	40.5
14	+1.33	3.0	5.00	7.50	10.0	12.50	17.25	23.75	30.75	41
15	+1.67	3.5	5.50	8.00	11.0	13.50	18.25	24.25	31.75	42
16	+2.00	4.0	6.00	8.75	11.5	14.50	18.75	25.50	32.75	43.5
17	+2.33	4.0	6.00	9.00	11.5	14.50	19.50	26.00	33.50	44.5
18	+2.67	4.5	6.25	9.25	12.0	15.00	20.50	26.50	34.25	45.5
Top minus bottom gaps		4.0	4.75	6.25	7.50	9.00	11.00	11.25	14.00	20.5
Top minus bottom raw scores		33.0	32.25	30.75	29.5	28.00	26.00	25.75	23.00	16.5

The raw scores of the target age (50–59) represent a perfect match between genes and environment. Hypothesis: at each age from 2 to 18, the gap between its raw scores and the target raw scores will not be uniform at all levels of achievement. As shown: (1) the gaps steadily increase by level of achievement from −2.67 SD below the median up to +2.67 SD above; (2) this is least apparent near the median (bold). For the derivation of this table, see Appendix II.

that seem virtually self-evident: that the culprit must be something that gives current environment independent potency; that can only be true if genes and current environment do not match perfectly at early ages; the obvious candidate is family environment, which is known to have an impact at early ages and fade away at about 18.

However, I should evidence my assumption that the factor at work is likely to be one rather than several. Therefore, I have done a correlation matrix of this sort: each age shows something at work that interferes with a perfect match between genes and current environment and whose magnitude progresses by level of achievement. Table 6 provides the matrix, which shows how highly correlated this something is between all ages; that is, age 2's progressive tendency has ben correlated with age 4's, with age 6's, and so forth; and the same has been done for all ages. Using this matrix, factor analysis shows that a single factor explains 99 per cent of the variance; the other 1 percent can be safely assigned to measurement error.

As to the identity of this single factor, some attempts at explanation alternative to my own can be dismissed quickly. First, the phenomenon cannot be a statistical artifact of how IQ tests are normed. The degree to which it is present varies greatly from one subtest to another. In the latest Wechsler Vocabulary data, like the Stanford-Binet, it is present although diminishing through the age of 18. For Arithmetic, as we shall see, you have to look back all the way to age 11.5 to find a similar potency. Since all subtests are normed the same, if it were a statistical artifact, there should be no subtest differences.

Second, recall that we have left chance environment out of our discussion. At all ages, fortuitous or un-fortuitous events occur that have nothing to do with the quality of genes or current environment and, therefore, have a potency that is independent of genes. However, chance events would affect our results only under one of two conditions:

(1) Chance events that affect cognitive abilities may be more likely at early ages or, conversely, more likely at adulthood. Divorced parents, moving from place to place, injury and

Table 6 Correlation matrix between the deviations from matched genes and environment: ages from 2 to 18 included; factor analysis indicates that a single factor is the cause

	2	4	6	8	10	12	14	16	18
2	—								
4	0.997	—							
6	0.998	0.999	—						
8	0.996	0.999	0.998	—					
10	0.995	0.998	0.998	0.998	—				
12	0.991	0.994	0.992	0.995	0.996	—			
14	0.996	0.995	0.996	0.995	0.996	0.992	—		
16	0.983	0.991	0.989	0.990	0.994	0.991	0.988	—	
18	0.973	0.969	0.970	0.975	0.973	0.975	0.984	0.961	—

Principle component analysis

	F1	F2	F3	F4	F5	F6	F7	F8	F9
% variance	99.066	0.552	0.226	0.083	0.037	0.015	0.010	0.008	0.003
Cumulative	99.066	99.618	99.884	99.927	99.964	99.979	99.989	99.997	100.0
Eigenvalue	8.916	0.050	0.020	0.007	0.003	0.001	0.001	0.001	0.000

illness, these traumas of childhood may be more or less likely than the adult traumas of marriage breakdown, unemployment, injury and illness. We know that this is false thanks to the twin studies: chance environment does not vary with age but approximates about 20 percent of IQ variance throughout life. How could this account for a phenomenon that disappears in early adulthood?

(2) Chance events may vary from early ages to adulthood in terms of the rate at which they occur according to performance level. Perhaps high achievers suffer more from chance events during childhood than high adult achievers, while low achievers do not. The fact that there is little variation in child/adult raw score gaps around the median suggests the contrary. We would have to assume that achievers in the vicinity of the median show no variation in the rate of chance events from childhood to adulthood, while other levels show a radical difference at the extremes. And again, this age differential would have to cease from early adulthood to the target age. And again, before it ceased it would have to vary from subtest to subtest.

Anything is logically possible, but in the absence of such evidence, I will set chance aside.

Measuring: raw scores into IQ scores

It is time to keep my promise to justify the IQ boons/ills of typical 17-year-olds at various levels – those levied by families above or below the percentile of one's achievement level. This entails converting raw score differences into IQ differences. This is purely mechanical and may bore scholars for whom it is an everyday task, while the general reader may find it taxing. Therefore, I have put the detail of the calculations in Box 1. Rather than converting scores at every level, I will do so at +2 SD (above the median), +1 SD, −1 SD, and −2 SD.

The box yields a pleasing symmetry. For Stanford-Binet (2001) Vocabulary, at age 11.5, those who perform at +2 SD above the median have been typically handicapped at +6.42 IQ points; at −2 SD below the median, they have benefitted by −6.43 IQ points. At +1 SD, they have been handicapped by +2.14 points; at −1 SD, they have benefitted by −4.29 points. Again, do not be confused by the signs. Remember a (+) is a fine that society collects to punish you for having a family whose cognitive value is worse than your level of performance; a (−) is a rebate society sends you to reward you for having a family whose cognitive value is better than your level of performance. My estimates of family effects for all mental abilities, whether based on Wechsler or Stanford-Binet subtests, have used this method.

Box 1   Converting raw score differences into IQ differences

I will use an example from Stanford-Binet (2001) Vocabulary. Tables (at every age) equate so many SDs above or below the median with raw scores. The median is set at an IQ of 100, and the size of the SD at 15. A raw score one SD above the median is by definition an IQ of 115, and a score one SD below the median is 85. You must compare the right ages. (1) The target age is the age at which family effects are gone and against which all earlier ages must be compared. I use the age at which raw score performance peaks. This will be defended in Chapter 5. Stanford-Binet (2001) Vocabulary has a target age of 50–59. (2) An appropriate earlier age is one with a raw score range that overlaps with that of age 50–59: I have chosen 11.5 years.

Explanation of calculations

In Step I, start with the +2 SD level of performance. The age 11.5 raw score has slipped well down the 50–59 scale. It is 46.5, which on the age 50–59 scale is only 2 points above the median.

The distance between the median and +1 SD is 7 points, so it is now only $\frac{2}{7}$ of an SD above the median. Therefore, the net loss from its original status of +2 SD is $1 + \frac{5}{7}$ SD. You know that an SD = 15, so simply multiply $1 + \frac{5}{7}$ SD (or 1.714 SD) by 15, which equals a gap of 25.71 IQ points. You do the same at all levels and find the usual "progressive pattern." The IQ gap between age 11.5 and age 50–59 falls away with level of performance: it is 25.71 IQ points at +2 SD; it is 12.86 IQ points at −2 SD!

Step II involves a trick. Remember that at the median there will be neither a family disadvantage nor a family advantage: those at the 50th percentile will come from as many above average families as below average. Therefore, the 11.5-year-olds at the median will on average have a perfect match between genes and current environment, despite the fact that there will be a mismatch at all other levels. In other words, the 11.5-years-olds at the median collectively have matching genes and environment at that age, just as the 50–59-year-olds at the median have matching genes and environment at that age (everyone at that age does). Therefore, the gap between the two ages must be purely one of age (growing maturity), and subtracting it from the gaps at all other levels will give a pure estimate of family effects.

Calculations: converting raw scores differences into IQ-point differences

Step I: norm age 11.5 on ages 50–59 – get the IQ gap at all standard score levels.

	11.5	50–59	11.5 normed on 54–59
+2 SD	**46.5**	58.5	Gap of $1 + \frac{5}{7}$ SD = 1.714 SD = 25.71 IQ points
+1 SD	41.5	51.5	Gap of $1 + \frac{3}{7}$ SD = 1.429 SD = 21.43 IQ points
		(46.5)	
Median	35.5	44.5	Gap of $1 + \frac{2}{7}$ SD = 1.286 SD = 19.29 IQ points
−1 SD	30.5	37.5	Gap of 1.000 SD = 15.00 IQ points
−2 SD	24.5	30.5	Gap of $\frac{6}{7}$ SD = 0.857 SD = 12.86 IQ points
−3 SD		(23.5)	

Step II: subtract the gap at the median from all other gaps to allow for maturity

---

+2 SD	25.71 − 19.29 = +6.42 IQ points (+ = family disadvantage)
+1 SD	21.43 − 19.29 = +2.14 IQ points (+ = family disadvantage)
Median	19.29 − 19.29 = 0.00 (by definition no disadvantage/advantage)
−1 SD	15.00 − 19.29 = −4.29 IQ points (− = family advantage)
−2 SD	12.86 − 19.29 = −6.43 IQ points (− = family advantage)

Figure 1 shows the overall results by age for Stanford-Binet (2001) Vocabulary. As you can see, above the median, we find *plus* IQ points which represent the greater score gaps between early ages and the target age thanks to the typical disadvantage from inferior families (at high levels of performance), and below the median we find *minus* values representing the typical advantage from superior families (at low levels of performance). Figure 1 demonstrates that family effects persist to a significant degree even past the age of 20, so the great debate is settled. Perhaps I should say 'half-settled'. As the next chapter shows, when we look at other cognitive abilities like arithmetic, family effects do disappear by the age of 17. So in a sense the negative side is not only half-wrong but also half-right.

A pause for consensus

My hope is that by now, you will be convinced that family is doing something to our data, that what it does declines with age, and that what it does can be measured. One of the new method's virtues is that it allows us to measure the decay of family effects

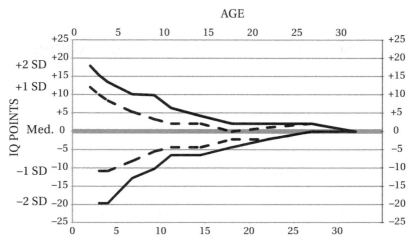

Figure 1 Stanford-Binet (2001) Vocabulary: decline of family effects with age at four performance levels. See Appendix II

by cognitive ability with precision. Unless I am mistaken, we have a rich new source of data, one that can supplement twin studies just by using the tables in test manuals, without the trouble and expense of elaborate kinship designs. I will call this new method the "Age-Table Method" thanks to the tables on which it is based.

Answers

(1) Age tables show that all the way from ages 2 to 18, a mismatch between genetic and family quality goes from benefitting low performers to penalizing high performers.

(2) These same tables equate raw scores with SDs, which allows us to convert raw score advantages or disadvantages into IQ points.

4   Slow and quick decay of family effects

Questions

(1) What cognitive abilities show the most persistent family effects and what cognitive abilities show the least?
(2) What seems to determine either persistence or lack of persistence?

I will anticipate the answers to our questions. The family has its most persistent effects on those cognitive abilities that children experience their parents using in everyday life: when they hear the language their parents use (Vocabulary and Similarities); when they hear the facts about the world their parents talk about (Information); when they observe their parent speaking and acting to cope with life (Understanding). Family effects are less persistent for those cognitive abilities that are "test-specific," abilities that are performed primarily in the test room. For example, Block Design and Object Assembly are rather like three-dimensional jigsaw puzzles; Picture Completion is spotting something missing in a picture (such as the hands of a clock).

Table 7a averages our total data (from 1950.5 to 2004.5) to contrast the persistence of family effects on Vocabulary (the most lasting effects) with Block Design and Picture Completion (among the subtests for which family effects fade quickly). At age 7, Vocabulary shows that family environment levies huge disadvantages/advantages, ranging from 6.92 IQ points to

Table 7a Averaged data. Contrasting slow decline of family environment with age (Vocabulary) with quick decline (Block Design and Picture Completion): IQ points at four performance levels; overall variance explained. See Tables 7b and 7c at the end of this chapter for other subtests

	Ages[a]						
	6.75/7	9.25/9.5	11.5/12	14.5	17.5/17	18–19	20–24
Wechsler Vocabulary (7)[b]							
+2 SD	+14.03	+8.82	+4.84	+2.31	−0.80	−0.64	−0.87
+1 SD	+10.85	+8.61	+6.17	+5.30	**+3.62**	**+2.84**	**+2.09**
−1 SD	−8.78	−6.99	−4.85	−1.74	**−2.89**	−1.72	−1.45
−2 SD	−20.75	−14.41	−11.47	−7.85	**−8.07**	**−5.41**	**−4.75**
Cor.	0.836	0.616	0.434	0.274	0.230	0.158	0.124
% var.	**69.89**	**37.93**	**18.84**	**7.51**	**5.29**	**2.49**	**1.54**
Stanford-Binet Vocabulary (more than 3)							
+2 SD	+14.84	+11.19	+7.81	+6.99	**+4.25**	—	—
+1 SD	+6.92	+4.42	+3.14	+2.97	+1.30	—	—
−1 SD	−10.48	−7.02	−5.26	−3.93	**−2.53**	—	—
−2 SD	−18.67	−15.51	−10.04	−8.71	**−5.57**	—	—
Cor.	0.771	0.559	0.391	0.333	0.197	—	—
% var.	**59.44**	**31.23**	**15.27**	**11.07**	**3.87**	—	—
Wechsler Block Design (3)							
+2 SD	+6.35	+6.39	+3.39	+2.92	+1.32	+1.65	—
+1 SD	+1.03	+1.62	+0.83	+0.53	−0.09	+0.24	—
−1 SD	−2.71	−0.44	−0.82	−0.31	**−2.84**	−1.64	—
−2 SD	−10.61	−3.27	−2.12	+0.32	**−2.27**	**−2.64**	—
Cor.	0.274	0.154	0.099	0.048	0.104	0.091	—
% var.	**7.51**	**2.38**	**0.98**	**0.23**	**1.08**	**0.83**	—
Wechsler Picture Completion (2)							
+2 SD	+7.68	+2.05	+3.13	−2.25	+1.88	−2.06	—
+1 SD	+4.26	+2.22	+1.36	−3.08	−1.66	−0.72	—
−1 SD	−2.73	−1.77	−0.52	−1.72	−1.66	**−2.87**	—
−2 SD	−3.80	−1.61	+0.87	−4.11	**−4.88**	−1.79	—
Cor.	0.288	0.132	0.068	0.011	0.072	0.047	—
% var.	**8.29**	**1.75**	**0.46**	**0.01**	**0.51**	**0.22**	—

[a]Wechsler results average four data sets (1950.5, 1975, 1992, and 2004.5).

[b]The numbers in brackets isolate values that are at least 2 IQ points at age 17 and above.

20.75 IQ points. Much diminished, these effects persist to age 17 and even up to age 24. Wechsler Vocabulary shows seven significant effects (more than 2 IQ points) at mature ages. Stanford-Binet Vocabulary is really the same in that no estimates are possible above age 17. On the other hand, the test-specific subtests show family effects of only 1.03 to 10.61 IQ points at age 7, and these have largely disappeared by age 14.5.

In this table and subsequent tables, certain values appear in italics. These values are contrary in sign – that is, you get some minus values rather than plus values above the median, and plus rather than minus values below. They may be measurement error. They are rare, usually quite small, and occur only at older ages.

I have now redeemed the first promissory note mentioned in Chapter 2. There I gave an estimate of the family disadvantages and advantages at various performance levels (as a prelude to coping with the SAT). It combined all of our six Vocabulary estimates. Above you have the four from Wechsler and the two from Stanford-Binet. If you combine them in a ratio of 2 to 1, you will replicate Table 1 in Chapter 2. Chapter 2 also promised to defend the estimates it used for the gaps between various percentiles of performance (say + 2 SD or the 98th) and the percentiles it posited for the typical cognitive value of the homes from which the subjects came (in this case the 69th). I have also used these gaps to calculate the *percentage of IQ variance* at various ages that family effects explain.

These estimates (% var.) are given in Table 7a. They are the most convenient way of estimating how long family effects persist until you reach the age at which they decline to almost nil. At that point, your current environment has swamped the influence of family and all that is left are the percentages of individual IQ differences that are due to genes, current environment matched to genes, and chance environment (now divided into pure chance and the exercise of your autonomy). Therefore, what we are about to do will redeem our second and last promissory note.

Partitioning IQ variance

How can we estimate the gap between those at various percentiles of performance and the percentiles of the cognitive quality of their families? As a preliminary, before the age at which the matching of an individual's genes with current environment begins (perhaps age 3 or 4), I posit that family determines virtually all of performance variance (this will be revisited in Chapter 6).

I will make assumptions about what determines the gaps between percentile of performance and percentile of home cognitive quality. Those who perform at the 98th percentile will come from few families that have a cognitive value in the bottom 30 percent (similarly those at −2 SD or the 2nd percentile will come from few families in the top 30 percent). Those who perform at the 84th percentile will come from few families in the bottom 15 percent (similarly those at −1 SD or the 16th percentile will come from families in the top 15 percent). These thresholds would not be literally true, of course. They are very much guesses about how much of a qualitative gap exists between various levels of performance and the family.

However, once you have made these assumptions, you can calculate a correlation coefficient between family effects and IQ, which in turn gives the percentage of IQ variance due to family effects. Box 2 explains how this can be done. It gives results for age 11.5 on Stanford-Binet Vocabulary: a correlation of 0.290 and 8.42 percent of variance explained. Table 7a reveals that my calculations give consistent results. Where the typical disadvantage/advantage of family effects give a large total, for example, 54.41 IQ points for Wechsler Vocabulary age 7, the percentage of variance they explain is also large (almost 70 percent). Where the total has faded away to a total of 13.78 IQ points (at age 17), the percentage of variance explained is about 5 percent.

In Chapter 6, we will compare all ages with the results of the twin studies. If they tally, that will show we are not too far off.

Box 2   Calculating correlation coefficients

At an early age, those who perform at 2 SD above the median (98th percentile) will open up a score gap with the target age in inverse proportion to the elite quality of the homes from which they come. For example, assume their homes were also at the 98th percentile in terms of quality. Then, even with a perfect correlation between family quality and IQ, no gap would be possible: the early age would effectively match the target age, at which +2 SD also shows an exact match between performance level and current environment (family having faded away to nil).

I have assumed that the homes at this level are elite but only to some degree: they consist of the upper 70 percent of home quality with the lower 30 percent missing, which means that, assuming a perfect correlation between family quality and IQ, they could open up a large proportion of the score gap of 30 IQ points. Thirty points is the maximum that separates +2 SD from the mean. As to how much their semi-elite character has lessened the possible score gap, we can use a table of values under a normal curve: the bottom 30 percent gone, the homes in question would average 0.4967 SD above the mean on the quality of homes curve. That amount is equivalent to a rise of 7.45 IQ points. Thus, their capacity to open up a gap with the target age is reduced by that value: 30.00 minus 7.45 = 22.55 IQ points as the maximum gap; always assuming a perfect correlation between family quality and IQ.

Therefore, to find the actual correlation between family quality and IQ, we have a simple equation: actual gap at +2 SD (divided by) 22.55 (equals) actual correlation. Similar calculations give an equation for +1 SD: actual gap (divided by) 10.89 (equals) actual correlation. Here only 15 percent of the home-quality curve is missing (equals 4.11 IQ points), and 15 − 4.11 = 10.89. Once you get the actual correlations, those squared give the percentage of IQ variance explained by family effects (again take the squaring on faith: it follows from the mathematics of a normal curve).

Calculations: converting typical IQ boons/ills into family percentage of IQ variance

For example, at age 11.5 on Stanford-Binet Vocabulary, the disadvantage at +2 SD is +6.42 IQ points. Dividing that by 22.55 gives 0.285 as the correlation between performance and family at that level. The disadvantage at +1 SD is +2.14 points. Dividing that by 10.89 gives 0.197 as the correlation at that level. Calculating the correlations at −1 SD and −2 SD uses the same divisors. In this case, −4.29 ÷ −10.89 = 0.394 as the correlation at −1 SD; and −6.43 ÷ −22.55 = 0.285 as the correlation at −2 SD.

If you average these four estimates of the correlation, you get 0.290 and that squared gives 8.42 percent of Stanford-Binet Vocabulary variance explained by family environment at age 11.5. Here are the calculations just described:

+2 SD: +6.42 ÷ 22.55 = 0.285	Average of correlations: 0.290
+1 SD: +2.14 ÷ 10.89 = 0.197	Square of 0.290 = 8.42 percent
−1 SD: − 4.29 ÷ 10.89 = 0.394	(of variance explained)
−2 SD: 6.43 ÷ 22.55 = 0.285	

Remember we are not seeking to get (and cannot hope to get) estimates accurate enough to second-guess the accuracy of how kinship studies partition IQ variance. My objective is only to get rough results that will tell us this: whether our estimates of how family effects trend downward with age are *similar* to the downward trend shown by the twin studies.

Post-2001 comparisons

I have contrasted the slow decay of family effects for Vocabulary with the quick decay for Block Design and Picture Completion, using the total data from 1950.5 to 2004.5. Table 8a does the same for the most recent data (post-2000).

Table 8a Post-2000 data. Contrasting slow decline of family environment with age (Vocabulary) with quick decline (Block Design and Picture Completion): IQ points at four performance levels; overall variance explained. See Tables 8b and 8c at the end of this chapter for other subtests

	Ages						
	6.75/7	9.25/9.5	11.5/12	14.5	17.5/17	18/18–19	20–24
Wechsler Vocabulary (9)							
+2 SD	+8.91	+4.45	+1.91	+0.75	+0.25	−2.25	−1.25
+1 SD	+9.52	+7.48	+5.77	+4.42	**+5.25**	**+4.75**	**+3.75**
−1 SD	−8.77	−6.23	−5.02	−1.68	**−4.18**	−4.11	−3.96
−2 SD	−15.25	−8.96	−7.75	−4.42	**−5.25**	**−5.00**	−4.50
Cor.	0.688	0.463	0.356	0.197	0.278	0.234	0.213
% var.	**47.30**	**21.45**	**12.67**	**3.89**	**7.71**	**5.47**	**4.54**
Stanford-Binet Vocabulary (more than 6)							
+2 SD	+10.17	+9.92	+6.42	+4.28	—	**+2.14**	+2.14
+1 SD	+5.39	+3.39	+2.14	+2.14	—	0.00	+1.07
−1 SD	−8.04	−5.44	−4.29	−4.29	—	**−2.14**	−2.15
−2 SD	−12.70	−10.18	−6.43	−6.43	—	**−4.29**	−2.15
Cor.	0.562	0.426	0.290	0.267	—	0.121	0.121
% var.	**31.57**	**18.13**	**8.42**	**7.12**	—	**1.46**	**1.46**
Wechsler Block Design (0)							
+2 SD	+5.22	−0.26	+0.73	+1.41	—	—	—
+1 SD	+2.88	+1.27	+0.12	+1.25	—	—	—
−1 SD	−4.07	−4.84	−1.55	+1.13	—	—	—
−2 SD	−13.98	−6.52	−4.65	+1.65	—	—	—
Cor.	0.372	0.210	0.098	0.000	—	—	—
% var.	**13.85**	**4.40**	**0.96**	**0.00**	—	—	—
Wechsler Picture Completion (0)							
+2 SD	+5.23	+3.18	0.00	−9.00	—	—	—
+1 SD	+1.14	+2.18	0.00	−6.50	—	—	—
−1 SD	−2.50	+0.68	−0.00	−3.55	—	—	—
−2 SD	−3.64	+0.79	+1.36	−3.55	—	—	—
Cor.	0.182	0.61	0.015	0.128	—	—	—
% var.	**3.31**	**0.37**	**0.02**	**1.64**	—	—	—

The numbers in brackets isolate values that amount to at least two IQ points at age 17 and above.

At age 7, Vocabulary shows that family environment still levies huge disadvantages/advantages ranging from 5.39 IQ points to 15.25 IQ points. Much diminished, these effects persist to age 17 and even up to age 24. Indeed, they are more persistent than the total data, what with Wechsler Vocabulary showing nine significant effects at mature ages. Stanford-Binet Vocabulary is not far behind, with six such effects. However, there is a difference in the variance explained by family at mature ages. For Wechsler Vocabulary, variance explained is still 7.71 percent at age 17, 5.47 percent at age 18, and 4.54 percent at ages 20–24. For the Stanford-Binet, the value for age 17 is missing but from the previous age, it looks like about 5 percent with only 1.46 percent thereafter.

On the other hand, the test-specific subtests show much smaller family effects: aside from one outlier, they range from only 1.14 to 5.23 IQ points at age 7. Moreover, compared to the full data, the recent data show that family effects disappear at an even earlier age. By 12 years, Block Design has already dropped to 0.96 percent of variance explained; by age 9.5 years, Picture Completions has already dropped to 0.37 percent. For the latter, family has virtually no persistence into the school years at all.

Comparing twelve cognitive abilities

All cognitive abilities (each IQ subtest measures its own ability) have their own distinctive age at which family effects diminish to the vanishing point. I will analyze eight Wechsler subtests on which performance can be traced from infancy to old age. Gale Roid helped me select four Stanford-Binet subtests similar enough to Wechsler subtests to afford comparative data. At the end of this chapter, I have attached Tables 7b, 7c, 8b, and 8c, so the reader can compare my discussion to the detailed results on the cognitive skills in question. They are the larger tables from which Table 7a and Table 8a in the text were taken. For background data and calculations that underlie these tables, see the appendices and simply match the appendix title to subtest title.

During this discussion, the reader may find it helpful to be reminded as to just what cognitive skills each subtest measures. Box 3 supplies that information.

Box 3   Description of subtests analyzed

Four Wechsler subtests with matching Stanford-Binet subtests

> Wechsler Vocabulary: What does "debilitating" mean?
>
> Stanford-Binet Verbal Knowledge: Also a vocabulary subtest
>
> Wechsler Arithmetic: If 4 toys cost 6 dollars, how much do 7 cost?
>
> Stanford-Binet Quantitative–Words: Like Wechsler Arithmetic
>
> Wechsler Comprehension: Why are streets usually numbered in order?
>
> Stanford-Binet Quantitative–Pictures (1985 only): Like Wechsler Comprehension
>
> Wechsler Information: On what continent is Argentina?
>
> Stanford-Binet Absurdities (1985 only): Like Wechsler Information
>
> Stanford-Binet Nonverbal Knowledge (2001 only): Like Wechsler Information

Four Wechsler subtests with no matching Stanford-Binet subtests

> Wechsler Similarities: In what way are "dogs" and "rabbits" alike?
>
> Wechsler Picture Completion: Indicate the missing part from an incomplete picture.

> Wechsler Block Design: Use blocks to replicate a two-color design.
>
> WISC Coding/WAIS Digit Symbol: Using a key, match symbols with shapes or numbers.

I will begin with the total data from 1950.5 to 2004.5. As expected, Wechsler and Stanford-Binet Vocabulary are highly comparable: 60 to 70 percent of variance explained by family at about age 7, falling away to 4 or 5 percent by age 17. Vocabulary is the only cognitive ability for which family effects persist to 24. Even after they start school, children and even teenagers (to some degree) listen to and speak to their parents. Wechsler Similarities (the ability to classify using general concepts) is interesting. The variance explained at age 7 (56 percent) is high, but more important, significant family effects number five at ages 17 and above. Similarities is a subtest that signals the ability of children to detach themselves from the concrete world and clothe it in abstractions, a prerequisite for coming to terms with formal schooling. Perhaps parental speech that has this character conditions the mind of the child.

At almost 75 percent, Wechsler Arithmetic shows the greatest variance explained by family at age 7. Whatever numeracy children have when they begin school depends on how much parents "drill" them (teach them how to add and subtract). Family effects show reasonable persistence, although this will disappear when we examine the most recent data. Parents are less influential (but still important) for Wechsler Information, a child's store of general knowledge, with variance explained at age 7 at 40 percent. Stanford-Binet believes that their Absurdities subtest (1985) and Nonverbal Knowledge subtest (2001) are analogous. Both SB subtests use picture absurdities that assume everyday information about wind, trees, airplanes, and laws of nature like gravity. The averaged data for the two does closely resemble Wechsler Information but this conceals the fact that Absurdities is much more family influenced than Nonverbal Knowledge.

As for the quick decay of family effects with age, Wechsler Comprehension is ambiguous. It has only three sizable disadvantages/advantages at age 17 or above. However, family effects explain 37 percent of variance at age 7. When we analyze post-2000 data, we will find that it moves up to join the slow-decaying subtests.

Given that Stanford-Binet Comprehension (last normed in 1985) seems to be a robust slow-decaying subtest, Comprehension probably qualifies as slow rather than fast.

As we have seen, Wechsler Block Design and Picture Completion are alike. Aside from showing minimal persistence at age 17 and above, family effects explain only 7 or 8 percent of the variance even at age 7. Wechsler Coding (the subtest is called Coding on the WISC but Digit Symbol on the WAIS) shows less persistence still. However, the percentage of variance explained at age 7 is moderately high at about 30 percent.

Post-2000 comparisons

As for post-2000 data, this consists of data from Wechsler WISC-IV/WAIS-IV, which average at being normed in 2004.5, and Stanford-Binet 5 (normed in 2001). Vocabulary explained at age 7 is down for both Stanford-Binet and Wechsler (to 30 and 50 percent), but recall that family effects have become even more persistent for ages 17 to 24. Wechsler Information rises to 60 percent of variance at early ages, and shows much the same kind of persistent effects save at the bottom levels. The supposedly analogous Stanford-Binet Nonverbal Knowledge (2001) show family weak at all ages. Perhaps its resemblance to Wechsler Information is suspect.

Wechsler Similarities now shows family explaining a huge 93 percent of variance at age 7. Although the percentage declines rapidly, family effects in early adulthood remain impressive. Arithmetic seems to have altered over time what with percentage of variance explained by family down to 0.46 percent by ages 14.5. Perhaps math teachers today match young children to a current environment closer to their genetic capacity than in the past. The Stanford-Binet subtest (Quantitative–Words) also uses arithmetical problems put verbally. It is similar in the sense that family effects are minimal (0.74 percent of variance explained) by age 14.5.

Stanford-Binet Comprehension has been included among recent results (despite the fact it was eliminated after 1985) because it is analogous to Wechsler Comprehension. It shows much more variance explained at age 7 (70 percent) than Wechsler Comprehension but its degree of persistence is similar. It has no values above age 17.5 only because the SB-4 (1985) subtest peaks at ages 20–24, which automatically becomes the target age. In the post-2000 data, the percentage for Coding rises to over 50 percent at age 7 but its family effects show little persistence.

Classification of subtests

I conclude that all twelve of our subtests fall into three classes. The first group consists of Vocabulary, Similarities, Information, and Comprehension. They have a mean of 52 percent of variance explained at age 7 for the averaged data (with Vocabulary top at 65 percent); and a mean of 53 percent in the post-2000 data (with Similarities top at 93 percent). All of them show persistence beyond age 17. These are cognitive skills that parents put on exhibit in everyday life. They speak in front of their children, using general terms to classify, display information, and explain the world to their children. Even after children begin school, they still have to communicate with their parents and still share their general awareness of the world.

The second group consists of Wechsler Block Design and Picture Completion (with the oddity of Stanford-Binet 5 Nonverbal Knowledge). These tests have a mean of about 8 percent of variance explained at age 7 and explain virtually no variance beyond age 12. Aside from the occasional jigsaw puzzle, they have no part in everyday life. Children never see their parents performing these cognitive tasks as part of normal behavior. Family effects are weak even among preschoolers. Since these subtests match environment with genetic potential so young, they would be an ideal measure (for, say, 5-year-olds) of genes for intelligence. They would, of course, sacrifice much external validity in terms of predicting significant behavior such as academic performance.

Finally, there is the pair of Arithmetic and Coding/Digit Symbol. Both show large percentages of variance explained at age 7, not only for averaged data but also for recent data: 30 and 50 percent for Coding, 74 and 79 percent for Arithmetic. As for the latter, whatever children know about numbers comes from parents before school; but at least in recent data, school quickly overwhelms family effects by offering children a current environment that matches the quality of their genetic promise.

Coding is more interesting. I hypothesize that for small children at least, character is a great advantage on this subtest: you have to persist at a boring task upon command with minimal distraction. Before school, parents inculcate the psychological traits of obedience and self-control. Once school begins, peers become a powerful influence on character, against which, as we know to our sorrow, parents fight a losing battle.

The text promised four tables as below.

Table 7b Averaged data. Slow decline of family environment with age on various subtests: points at four performance levels; overall variance explained

	Ages[a]						
	6.75/7	9.25/9.5	11.5/12	14.5	17.5/17	18–19	20–24
Wechsler Vocabulary (7)[b]							
+2 SD	+14.03	+8.82	+4.84	+2.31	−0.80	−0.64	−0.87
+1 SD	+10.85	+8.61	+6.17	+5.30	**+3.62**	**+2.84**	**+2.09**
−1 SD	−8.78	−6.99	−4.85	−1.74	**−2.89**	−1.72	−1.45
−2 SD	−20.75	−14.41	−11.47	−7.85	**−8.07**	**−5.41**	**−4.75**
Cor.	0.836	0.616	0.434	0.274	0.230	0.158	0.124
% var.	**69.89**	**37.93**	**18.84**	**7.51**	**5.29**	**2.49**	**1.54**
Stanford-Binet Vocabulary (more than 3)							
+2 SD	+14.84	+11.19	+7.81	+6.99	**+4.25**	—	—
+1 SD	+6.92	+4.42	+3.14	+2.97	+1.30	—	—
−1 SD	−10.48	−7.02	−5.26	−3.93	**−2.53**	—	—
−2 SD	−18.67	−15.51	−10.04	−8.71	**−5.57**	—	—
Cor.	0.771	0.559	0.391	0.333	0.197	—	—
% var.	**59.44**	**31.23**	**15.27**	**11.07**	**3.87**	—	—
Wechsler Similarities (5)							
+2 SD	+7.81	+4.39	+1.92	+0.93	+1.69	+0.24	−1.28
+1 SD	+8.56	+4.73	+4.07	+3.05	**+3.85**	**+2.03**	+1.27
−1 SD	−10.75	−6.43	−4.41	−1.23	**−2.67**	−1.77	**−2.25**
−2 SD	−19.48	−13.57	−7.26	−1.50	**−2.25**	−1.05	−0.38
Cor.	0.746	0.455	0.296	0.125	0.194	0.102	0.071
% var.	**55.63**	**20.69**	**8.77**	**1.57**	**3.75**	**1.04**	**0.50**
Wechsler Arithmetic (5)							
+2 SD	+19.45	+7.96	+4.21	+2.45	**+2.33**	−0.29	+0.25
+1 SD	+11.16	+5.15	+3.27	+1.22	**+2.33**	−0.28	−0.63
−1 SD	−10.37	−8.12	−4.98	−3.84	**−2.62**	−1.33	+0.35
−2 SD	−13.36	−12.62	−5.26	−4.01	**−3.28**	**−2.35**	−0.70
Corr.	0.862	0.533	0.294	0.188	0.176	0.047	0.012
% var.	**74.33**	**28.41**	**8.65**	**3.53**	**3.09**	**0.22**	**0.01**

(*continued*)

Table 7b Averaged data. Slow decline of family environment with age on various subtests: points at four performance levels; overall variance explained (*continued*)

Wechsler Information (4)						
+2 SD	+12.87	+13.51	+8.27	+3.66	**+3.31**	**+2.55**	+0.77
+1 SD	+7.32	+8.58	+4.89	+3.95	**+3.06**	**+2.06**	+0.52
−1 SD	−6.38	−2.17	−0.80	−1.77	−0.12	*+0.50*	−0.39
−2 SD	−16.04	−8.69	−5.15	−3.29	−1.89	−1.32	−0.45
Cor.	0.635	0.493	0.279	0.209	0.131	0.079	0.035
% var.	**40.31**	**24.29**	**7.79**	**4.36**	**1.71**	**0.62**	**0.12**

Stanford-Binet like Information (Absurdities + Nonverbal Knowledge)[c]						
+2 SD	+14.63	+10.40	+6.66	+2.10	—	—	—
+1 SD	+7.74	+6.01	+3.20	+2.60	—	—	—
−1 SD	−8.48	−8.16	−4.39	−2.08	—	—	—
−2 SD	−13.37	−15.10	−8.59	−4.27	—	—	—
Cor.	0.683	0.608	0.343	0.178	—	—	—
% var.	**46.65**	**36.98**	**11.76**	**3.17**	—	—	—

[a]The first age given is from the Stanford-Binet, the second from the Wechsler.

[b]The numbers in brackets stand for the number of values that are still at least 2 IQ points at age 17 and above.

[c]Averaging Stanford-Binet Absurdities (1985) and Stanford-Binet Nonverbal Knowledge (2001) obscures the fact that the latter showed weak effects of family environment (see Table 9 below).

Table 7c Averaged data.[a] Quick decline of family environment with age on various subtests: points at four performance levels; overall variance explained

				Ages			
	7	9.5	12	14.5	17	18–19	20–24
Wechsler Comprehension (3)[b]							
+2 SD	+8.81	+3.69	+0.94	−0.51	−0.78	−1.02	−0.09
+1 SD	+7.46	+6.50	+3.58	+1.34	**+2.71**	+1.43	+1.32
−1 SD	−8.53	−3.56	−1.40	−0.29	−1.91	−1.10	+0.36
−2 SD	−13.05	−8.95	−4.49	−3.36	**−3.49**	**−2.75**	−0.72
Cor.	0.610	0.371	0.174	0.069	0.136	0.077	0.029
% Var.	**37.17**	**13.77**	**3.06**	**0.48**	**1.85**	**0.59**	**0.08**
Wechsler Block Design (3)							
+2 SD	+6.35	+6.39	+3.39	+2.92	+1.32	+1.65	—
+1 SD	+1.03	+1.62	+0.83	+0.53	−0.09	+0.24	—
−1 SD	−2.71	−0.44	−0.82	−0.31	**−2.84**	−1.64	—
−2 SD	−10.61	−3.27	−2.12	+0.32	**−2.27**	**−2.64**	—
Cor.	0.274	0.154	0.099	0.048	0.104	0.091	—
% Var.	**7.51**	**2.38**	**0.98**	**0.23**	**1.08**	**0.83**	—
Wechsler Picture Completion (2)							
+2 SD	+7.68	+2.05	+3.13	−2.25	+1.88	−2.06	—
+1 SD	+4.26	+2.22	+1.36	−3.08	−1.66	−0.72	—
−1 SD	−2.73	−1.77	−0.52	−1.72	−1.66	**−2.87**	—
−2 SD	−3.80	−1.61	+0.87	−4.11	**−4.88**	−1.79	—
Cor.	0.288	0.132	0.068	0.011	0.072	0.047	—
% Var	**8.29**	**1.75**	**0.46**	**0.01**	**0.51**	**0.22**	—
Wechsler Coding/Digit Symbol (2)							
+2 SD	—	+10.97	+6.75	+2.41	+0.03	+0.65	—
+1 SD	—	+6.22	+3.94	+2.12	−0.12	+0.72	—
−1 SD	—	−6.31	−4.12	−2.06	−0.67	−1.27	—
−2 SD	—	−13.75	−8.07	−4.75	**−2.21**	**−2.15**	—
Cor.	—	0.560	0.349	0.176	0.038	0.077	—
% Var.	—	**31.33**	**12.19**	**3.09**	**0.14**	**0.59**	—

[a]Wechsler results average four data sets (1950.5, 1975, 1992, and 2004.5).

[b]The numbers in brackets isolate values that are at least two IQ points at age 17 and above.

Table 8b Post-2000 data. Slow decline of family environment with age on various subtests: points at four performance levels; overall variance explained

	Ages						
	6.75/7	9.25/9.5	11.5/12	14.5	17.5/17	18/18–19	20–24
Wechsler Vocabulary (9)							
+2 SD	+8.91	+4.45	+1.91	+0.75	+0.25	−2.25	−1.25
+1 SD	+9.52	+7.48	+5.77	+4.42	**+5.25**	**+4.75**	**+3.75**
−1 SD	−8.77	−6.23	−5.02	−1.68	**−4.18**	**−4.11**	**−3.96**
−2 SD	−15.25	−8.96	−7.75	−4.42	**−5.25**	**−5.00**	**−4.50**
Cor.	0.688	0.463	0.356	0.197	0.278	0.234	0.213
% var.	**47.30**	**21.45**	**12.67**	**3.89**	**7.71**	**5.47**	**4.54**
Stanford-Binet Vocabulary (more than 6)							
+2 SD	+10.17	+9.92	+6.42	+4.28	—	**+2.14**	**+2.14**
+1 SD	+5.39	+3.39	+2.14	+2.14	—	0.00	+1.07
−1 SD	−8.04	−5.44	−4.29	−4.29	—	**−2.14**	**−2.15**
−2 SD	−12.70	−10.18	−6.43	−6.43	—	**−4.29**	**−2.15**
Cor.	0.562	0.426	0.290	0.267	—	0.121	0.121
% var.	**31.57**	**18.13**	**8.42**	**7.12**	—	**1.46**	**1.46**
Wechsler Information (9)							
+2 SD	+16.90	+16.90	+11.90	+6.75	**+3.57**	**+3.57**	**+3.57**
+1 SD	+14.09	+14.09	+9.09	+8.18	**+5.00**	**+5.00**	**+3.33**
−1 SD	−2.00	−2.00	−2.00	−3.82	**−2.00**	**−2.00**	**−2.00**
−2 SD	−18.42	−19.37	−7.17	+2.68	−0.50	−0.50	−0.50
Cor.	0.761	0.775	0.466	0.321	0.206	0.206	0.168
% var.	**57.91**	**60.04**	**21.73**	**10.29**	**4.24**	**4.24**	**2.81**
Stanford-Binet like Information (Nonverbal Knowledge)[a]							
+2 SD	0.00	+2.50	0.00	−1.00	—	−2.00	—
+1 SD	0.00	+2.50	0.00	0.00	—	0.00	—
−1 SD	−6.25	−3.75	−1.25	0.00	—	0.00	—
−2 SD	−10.00	−7.50	−5.00	−2.51	—	−1.25	—
Cor.	0.254	0.253	0.084	0.017	—	0.009	—
% var.	**6.46**	**6.39**	**0.71**	**0.03**	—	**0.00**	—

	Wechsler Similarities (8)						
+2 SD	+7.83	+2.97	+3.42	−0.25	+1.25	−0.83	−1.25
+1 SD	+10.33	+4.67	+5.92	+5.00	+5.00	+2.38	+1.43
−1 SD	−16.12	−11.34	−9.28	−5.62	−3.75	−4.58	−2.50
−2 SD	−24.53	−14.56	−10.19	−5.84	−3.08	−3.71	−2.69
Cor.	0.966	0.562	0.500	0.306	0.249	0.192	0.106
% var.	93.32	31.58	25.00	9.36	6.19	3.69	1.13
	Wechsler Comprehension (7)						
+2 SD	+5.48	+3.81	+2.75	+3.15	+1.25	−2.50	+0.83
+1 SD	+6.23	+4.73	+4.73	+3.33	+5.00	+1.25	+0.83
−1 SD	−4.80	−2.34	−2.07	−2.74	−3.57	−4.64	−2.18
−2 SD	−7.53	−3.43	−2.07	−2.74	−3.57	−4.64	−2.18
Cor.	0.398	0.243	0.210	0.205	0.250	0.159	0.103
% var.	15.82	5.89	4.40	4.20	6.25	2.53	1.06
	Stanford-Binet like Comprehension (Quantitative–Pictures) (more than 2)[b]						
+2 SD	+22.74	+14.39	+8.57	+5.61	+3.48	—	—
+1 SD	+10.21	+8.28	+4.29	+.39	+3.75	—	—
−1 SD	−6.58	−8.06	−3.88	−0.81	+0.36	—	—
−2 SD	−17.83	−17.33	−10.44	−4.11	−0.26	—	—
Cor.	0.835	0.727	0.398	0.190	0.120		
% var.	69.74	52.83	15.85	3.61	1.44		

[a]Stanford-Binet Nonverbal Knowledge has been included in this table because of its supposed similarity to Wechsler Information. Its results concerning the persistence of family effects are *not* similar (weak effects).

[b]Stanford-Binet Comprehension, despite dating from 1985, has been included in this table because it is the most recent Stanford-Binet data.

Table 8c Post-2000 data. Quick decline of family environment with age on various subtests: points at four performance levels; overall variance explained

	6.75/7	9.25/9.5	11.5/12	14.5	17	18/18–19	20–24
				Ages			
Wechsler Arithmetic (1)[a]							
+2 SD	+15.79	+3.21	+3.21	−0.54	−0.54	−2.14	−2.14
+1 SD	+1.02	+4.46	+4.46	+1.25	+0.71	−2.14	−2.14
−1 SD	−11.54	−10.05	−4.97	−2.93	−0.54	−0.26	−0.26
−2 SD	−15.53	−7.49	−0.54	+1.96	**−2.38**	+1.19	+1.19
Cor.	0.865	0.452	0.258	0.068	0.049	0.080	0.080
% var.	**78.84**	**20.42**	**6.66**	**0.46**	**0.24**	**0.64**	**0.64**
Stanford-Binet like Arithmetic (Quantitative–Words) (1)							
+2 SD	+10.00	+5.00	+5.00	+1.25	—	**+5.00**	—
+1 SD	+5.00	+2.50	+5.00	0.00	—	0.00	—
−1 SD	−7.14	−4.64	−2.14	−1.43	—	+4.29	—
−2 SD	−9.29	−9.29	−4.69	−3.57	—	+2.14	—
Cor.	0.493	0.323	0.267	0.086	—	0.067	—
% var.	**24.28**	**10.42**	**7.13**	**0.74**	—	**0.48**	—
Wechsler Coding/Digit Symbol							
+2 SD	+15.67	+11.23	+8.14	+0.74	—	—	—
+1 SD	+8.06	+5.35	+3.49	+2.45	—	—	—
−1 SD	−8.70	−7.40	−4.14	−1.71	—	—	—
−2 SD	−13.76	−12.45	−7.24	−4.54	—	—	—
Cor.	0.711	0.555	0.346	0.154	—	—	—
% var.	**50.55**	**30.82**	**11.95**	**2.37**	—	—	—
Wechsler Block Design (0)							
+2 SD	+5.22	−0.26	+0.73	+1.41	—	—	—
+1 SD	+2.88	+1.27	+0.12	+1.25	—	—	—
−1 SD	−4.07	−4.84	−1.55	+1.13	—	—	—
−2 SD	−13.98	−6.52	−4.65	+1.65	—	—	—
Cor.	0.372	0.210	0.098	0.000	—	—	—
% var.	**13.85**	**4.40**	**0.96**	**0.00**	—	—	—

Wechsler Picture Completion (o)						
+2 SD	+5.23	+3.18	0.00	−9.00	—	—	—
+1 SD	+1.14	+2.18	0.00	−6.50	—	—	—
−1 SD	−2.50	+0.68	−0.00	−3.55	—	—	—
−2 SD	−3.64	+0.79	+1.36	−3.55	—	—	—
Cor.	0.182	0.61	0.015	0.128	—	—	—
% var.	3.31	0.37	0.02	1.64	—	—	—

[a]The numbers in brackets stand for the number of values that are at least 2 IQ points at age 17 and above. The absence of values for most subtests is due to the fact that the target age (after which aging lowers scores) is 17.

Answers

The answers to our questions have been anticipated and therefore, I will be brief:

(1) The cognitive abilities that show the most persistent family effects are Vocabulary, Similarities, Information, and Comprehension. Those that show the least are Wechsler Block Design and Picture Completion. In the most recent data, Arithmetic and Coding/Digit Symbol show large percentages of variance that are explained at age 7 but that fade away sometime between ages 11.5 and 14.5.

(2) The persistent abilities benefit from the fact that parents put them on exhibit in everyday life. The least persistent suffer because they have no part in everyday life. A few, like Arithmetic and Coding/Digit Symbol, are special cases.

5 Reconciliation with twins and adoptions

Questions

(1) Are my estimates of the cognitive variance due to family effects compatible with the estimates of "common environment" from kinship studies?
(2) Do the IQ gains of adopted children confirm the kinship studies?

My estimates of the percentage of IQ variance family accounts for will prove sufficient to accomplish their purpose. I aim at *supplementing* kinship studies (with fresh estimates of family effects by performance level and subtest), not at *replacing* them. This chapter will show that my estimates are comparable with the twin studies, taking the twin estimates as a given. It will also show that my estimates are in accord with adoption studies, with particular reference to identifying the age at which family effects cease.

Twin studies

We can use my results for the percentage of variance due to family effects to partition cognitive variance into its three main components. These are: genes (including environment matched to genes), family, and current environment uncorrelated with genes. Fortunately, kinship studies show that "uncorrelated with genes" environment (or chance environment or uncommon environment) is steady between ages 6 and adulthood, which is something we would expect from a set of "random" factors. If we add that percentage on to the family percentage and deduct the sum from 100 percent, we get my own estimate of the influence of genes.

Dutch kinship study

Using Stanford-Binet (2001) Vocabulary, Table 9 performs that service. The Dutch values are from Holland (McGue et al., 1993) and their estimates are typical of kinship studies.

Their estimate of 18 percent accounted for by chance ("uncommon environment") is a bit lower than the usual 20 percent (Haworth et al., 2010; McGue et al., 1993), but it is close enough. The values in bold show that we have achieved our objective of a good match with kinship data: the six comparisons show on average a difference of only 7.45 percent; the match is almost perfect at age 18.

Table 9 Comparison of genetic proportion of variance (h^2) between Stanford-Binet 5 Vocabulary (2001) and Dutch kinship estimates

My ages	Ave. cor. by age	% var. family	% var. uncommon	% var. genes (mine)	% var. genes (Dutch)	Dutch ages
4	0.806	64.96	18.00	**17.04**	**22**	5
6.75	0.562	31.57	18.00	**50.43**	**40**	7
9.25	0.426	18.13	18.00	**63.87**	**54**	10
11.5	0.290	8.42	18.00	**73.58**	**85**	12
14.5	0.267	7.12	18.00	**74.88**	—	—
18	0.121	1.46	18.00	**80.54**	**82**	18
20–24	0.121	1.46	18.00	**80.54**	—	—
25–29	0.073	0.53	18.00	**81.47**	**88**	26

The termination of family effects

However, there is one result of the twin studies we have yet to confirm: that family effects disappear entirely sometime during adulthood. The interesting subtests are those for which the peak age is a mature one and for which family effects are the most persistent. Using the most recent data, the subtests in question are Wechsler Vocabulary, Information, Comprehension, and Similarities. First, I will extend the method's analysis of family effects well into the adult years (see the relevant appendices for these subtests); second, I will argue that a pattern emerges that makes it very likely that family effects are exhausted at the target age.

In Table 10, the method shows that Vocabulary does have significant family effects (two points or more) particularly below the mean both at ages 25–29 and 30–34.

However, at ages 35–44, values are near randomly distributed (indeed, there are "negative" family effects above the mean) and only 0.04 percent of variance is explained. Information shows an almost identical pattern. Comprehension shows very significant effects below the mean (fully 4.53 points) but these are largely gone by age 30. Similarities show no really significant effects even at age 25–29. The values that indicate positive family effects are balanced by other values that indicate negative effects (remember, you want pluses above the median and minuses below the median). The total variance explained at age 25–29 is less than 1 percent.

Unfortunately, since Table 10 is based on the assumption that family effects fade away by target ages not much older than 35–44, it begs the question. How do we know that family effects do not persist at older ages, and go on persisting until death? I will call this the rock hypothesis: that at a certain age the erosion of family effects ceases and that the effects that remain are impervious to change. Clearly, our method cannot conclusively falsify such

Table 10 Persistence of family effects into adulthood (WAIS-IV data: 2007)

	Ages		
	25–29	30–34	35–44
Vocabulary (5)			
+2 SD	−2.00	−1.75	−1.50
+1 SD	**+2.00**	+0.25	−0.50
−1 SD	**−3.14**	**−2.32**	−1.50
−2 SD	**−3.50**	**−2.50**	−1.50
Correlation	0.127	0.062	0.039
% var.	**1.62**	**0.38**	**0.04**
Information (4)			
+2 SD	+3.57	−0.71	−0.71
+1 SD	**+3.33**	**+3.33**	+1.67
−1 SD	**−2.00**	**−2.00**	−1.83
−2 SD	−0.50	−0.50	−0.33
Correlation	0.168	0.120	0.076
% var.	**2.81**	**1.44**	**0.58**
Comprehension (2)			
+2 SD	−2.92	+0.42	−0.71
+1 SD	+0.83	+0.42	+1.67
−1 SD	**−4.53**	−1.09	0.00
−2 SD	**−4.53**	−1.09	0.00
Correlation	0.141	0.052	0.046
% var.	**1.99**	**0.27**	**0.22**
Similarities (1)			
+2 SD	−2.49	−1.67	—
+1 SD	+0.95	+0.47	—
−1 SD	**−3.33**	−1.67	—
−2 SD	−1.02	+0.64	—
Correlation	0.082	0.024	—
% var.	**0.67**	**0.06**	—

Note: the target age for Similarities is 35–44, and therefore the method's estimate of the persistence of family effects must stop at latest age younger than that.

a hypothesis, but let us state its assumptions and assess whether they are likely.

Turning to Table 11, the rock hypothesis is applied to the three most promising subtests. It posits that significant family effects remain at age 45 and presumably thereafter, and thus can explain 10 percent of performance variance. The rows that operationalize the rock hypothesis simply add 10 percent of variance explained to what our data show at adult ages (for example, Vocabulary ages 20–25: our data 1.62 percent; rock hypothesis becomes 11.62 percent).

Our data is labeled the erosion hypothesis. Note that at age 12, when our data explained 10 percent or more of variance, that percentage was highly vulnerable: within 15 years, it dwindled to almost nothing as family was swamped by current environment. But at age 25, when the rock hypothesis scenario posits 10 percent or more, that same percentage must be assumed somehow to have become invulnerable. This is despite the fact that the subjects are

Table 11 Whether family environment explains significant variance at age 45: rock hypothesis versus erosion hypothesis (WAIS-IV data: 2007)

	Ages				
	12	25–29	30–34	35–44	45
	Vocabulary				
% var. rock	—	11.62	10.38	10.04	10.00
% var. erosion	12.67	1.62	0.38	0.04	—
	Information				
% var. rock	—	12.81	11.44	10.58	10.00
% var. erosion	21.73	2.81	1.44	0.58	—
	Similarities				
% var. rock	—	10.67	10.06	—	10.00
% var. erosion	25.00	0.67	0.06	—	—

some 15 years older and, presumably, their family environment should be feebler (being further into the past).

It could be argued that current environment has somehow been robbed of its potency to swamp past environments: that at some age (for some unstated reason) current environment is no longer "self-contained" in its impact on current performance; rather, suddenly, its effects have to be averaged with what previous current environments dictated about IQ. Such a thing, like all things, is logically possible. But I remain skeptical until some plausible explanation for this radical shift is put.

Having shown that there is little wiggle-room for those who believe family effects never fade, what of those like Jensen (1998) who claim that they are entirely gone by age 17? The method gives them no latitude. They may declare arbitrarily what age we set as the target age. But if they posit that there is a perfect match between genes and current environment at 17, the method then shows that at all ages above 17 there is a mismatch of a new and peculiar sort. All those who perform above the median are favored by some environmental factor not correlated with genes, and all those below the median are disadvantaged thereby. In other words having died at 17, family effects suddenly reappear with the *opposite* effects they had before.

In sum, the naysayers can either posit a perfect match at age 50 and concede a mismatch at 17, or can posit a match at age 17 and endorse an absurdity.

Adoption studies

Some adoption studies appear to confirm that family effects on cognition are gone by age 17 but, in my opinion, their samples do not include adoptive homes of low cognitive quality. On the other hand, scholars like Dick Nisbett have located studies that yield large family effects and make him question the twin literature. Up to now, as far as I am aware, there is only one adoption study in which subjects were tested at near maturity (ages 18–20) and the adoption effects were small. However, much larger effects were found from ages 12 to 14, which may make it seem implausible that all benefits would disappear a few years later.

Nisbett and Kendler

The literature yields four important studies, three from France analyzed by Nisbett (2009) and one from Sweden reported by Kendler et al. (2015).

(1) Schiff et al. (1978) compared French children adopted by upper-middle-class families with their siblings who remained in a lower-class environment. Depending on the test, they profited by 11.5 to 16.1 IQ points. They were abandoned (and adopted) between 1962 and 1969; the data was presumably collected in 1977; therefore the median age at testing would be about 11 or 12.

(2) Duyme (1981) studied abused children who were adopted and tested at age 4.5 (IQ range 61–85) and retested at age 14. All made IQ gains but adoptees into homes of high socioeconomic status (SES) gained 12 more points than those adopted by poor, unskilled workers.

(3) Capron and Duyme (1989) studied adoptees who were born to either lower- or upper-class parents and raised in either lower- or upper-class homes. Comparing professional or top-management homes (sixteen years of education: very high in the France of 1988) and the homes of unskilled or semiskilled workers (nine years of education or less) gave an IQ difference of 11.65 points. The mean age of testing was 14 years and they used the WISC-R.

(4) Kendler et al. (2015) analyzed data from Swedish military tests that include every male aged 18–20 except foreigners and those severely disabled. They calculated IQs from the four tests given, which measured verbal, logical, spatial, and technical abilities. They identified 436 subjects who were full siblings (same father and mother) and among whom one was adopted away and one home reared. The adoptees had a mean IQ 4.41 points above that of the home reared. They also identified 2,341 male half-siblings (one birth parent in common), and the adoptees had an advantage of 3.18 IQ points. I will take the latter as representing a larger sample. There is no reason to believe that the child of superior genetic quality was adopted in either sample.

All of the French studies compared homes that were radically different. If we put the elite homes at the 84th percentile of cognitive quality and the substandard homes at the 16th percentile, they were 2 SD apart. The Swedish study did not isolate homes that dramatically contrasted in quality. Their method yielded whatever quality difference normally separates adoptive and non-adoptive homes. Based on years of parental education, the two were 0.563 SD apart.

Reconciliation

Our own data affords correlations between family quality and IQ at various ages. These tell us the number of IQ points that a gap of so many SD (of cognitive quality) between two homes entails. For example, if two homes are two SD apart that is the equivalent of 30 IQ points and if the correlation were perfect (= 1.000), 30 IQ points would be the difference. But if the correlation were half that (= 0.500) you would take 0.500 times 30 to get the true difference in IQ points, or 15 points.

From the latest data (Wechsler 2004.5), I selected the four subtests that are most predictive of Full-Scale IQ – namely, Vocabulary, Information, Similarities, and Comprehension – and averaged the correlations. Putting the French homes at 2 SD apart for cognitive quality, the average correlation times 30 (= 2 SD) predicts the IQ difference. I did this for ages 11.5 and 14 to match the ages of the French data; and for age 17.5 as well. As for the Swedish data, the tests they used come closest to Wechsler Vocabulary plus Raven's Progressive Matrices. As we shall see, the persistence of family effects for these two tests are almost identical (Table 15). You must allow for the fact that the cognitive difference between adoptive and non-adoptive homes in Sweden was only 0.563 SD. This means multiplying your correlation times 0.563 of an SD, which equals multiplying your correlation times 8.445 points (0.563 × 15 = 8.445).

The result is Table 12. The key comparative values are in bold. For age 11.5, my results (11.49 IQ points) come very close to predicting the range of the French adoption results (11.50–16.10). For age 14, my prediction (7.72 IQ points) falls short of the French results (11.65–12.00). Perhaps French families circa 1980 had more persistent effects than US families did soon after the dawn of the twenty-first century. However, my prediction for age 17.5 at 7.37 IQ points would probably be much closer to adoption results, although at present none exist. The prediction for 18–20 year olds in Sweden is low by about a point (my 1.89 as opposed to its 3.18).

Recall that if I have overestimated the gap between the percentiles of levels of performance and the percentiles of the cognitive values of homes, all correlations would rise. Accordingly, the predictions of adoption effects and the adoption results would match very closely. Still, the predictions are pretty good. If I have shown that my estimates tally with both the results of the twin studies and the results of the adoption studies, then twins and adoptions have been reconciled.

Table 12 The Age-Table Method's prediction of adoption effects and the results of adoption studies compared

	Age 11.5			
	Vocabulary	Information	Similarities	Comprehension
Correlation	0.356	0.466	0.500	0.210
C × 30 points	10.68	13.98	15.00	6.30
Ave. four tests	**11.49**	(predicted effects)		
Adoption	**11.50–16.10**	(actual effects)		
	Age 14			
	Vocabulary	Information	Similarities	Comprehension
Correlation	0.197	0.321	0.306	0.205
C × 30 points	5.91	9.63	9.18	6.15
Ave. four tests	**7.72**	(predicted effects)		
Adoption	**11.65–12.00**	(actual effects)		
	Age 17.5			
	Vocabulary	Information	Similarities	Comprehension
Correlation	0.278	0.206	0.249	0.250
C × 30 points	8.34	6.18	7.47	7.50
Ave. four tests	**7.37**	(predicted effects)		
Adoption	**NIL**	(no data age 17.5)		
	Ages 18–20			
	Vocabulary	Raven's		
Correlation	0.224	(0.224)		
C × 8.445 points	**1.89**	(1.89)	(predicted effects)	
Adoption	**3.18**	(actual effects)		

Calculations:

Ages 11.5, 14, and 17.5: Use the values for age 11.5 and 14.5 from Table 7c. Take the correlation × 30 IQ points (as = 2 SD).

Ages 18–20: Use the values for ages 18 and 20–24 from Table 7c. Average the two correlations (0.234 and 0.213) to get the correlation for "18–20". Wechsler Vocabulary and Raven's results are presumed the same. The reason the result is multiplied by 8.445 points is that this represents a difference of 0.563 SD and 15.00 × 0.563 = 8.445 IQ points.

Answers

(1) After cognitive variance has been partitioned into genes, common environment, and uncommon environment, the results of our method and kinship studies are highly similar. Our method by its very nature cannot confirm that family effects entirely disappear during adulthood, but it suggests that this is so.

(2) The Age-Table Method also shows that the IQ gains of adopted children are compatible with the family portion of IQ variance that the twin studies yield. There is little direct evidence, but adoption effects probably do exist at age 18–20. However, they are low enough to suggest that they do not persist much after that.

6   The fairness factor

Questions

(1) Do parents attempt to give all of their children roughly the same quality of cognitive environment?

(2) Which cognitive abilities are most influenced by their efforts?

The Age-Table Method provides some evidence, which, as far as I know, is novel. It measures the extent to which parents try to give their preschool children environments of equal cognitive quality, despite the fact that their children differ in giftedness.

Both the Stanford-Binet and the Wechsler tests (thanks to the Wechsler Pre-Primary Scale of Intelligence or WPPSI) allow us to partition variance among very young children who are still almost entirely conditioned by the home environment their parents provide. Given the high values for family effects at age 7, Vocabulary already at 65 percent and Similarities verging on complete dominance, we would anticipate that family would overwhelm genes among younger children. Bouchard (2013) was correct when he speculated that the variance explained by genes among preschoolers would be minimal. When the brain begins to cognize, family effects may be as high as 80 percent, with the rest split between genes and uncorrelated environment (chance factors like infant illness). Indeed, genes probably account for less than 10 percent.

This means that the direct effects of genes on individual differences in brain physiology may be slight. This fact has implications for scans of the infant brain (magnetic resonance), assuming we want to predict which brain has the most genetic potential. It would be as difficult as looking at two healthy seeds under a microscope and predicting which would grow into the taller plant. This in no way dismisses the potency of whatever genetic differences exist. However subtle they may be at birth, during the course of any normal life they are capable of matching quality of environment so as to produce a combination that differentiates all the levels of cognitive achievement we see around us.

The evidence

When we apply the method to preschool data, we find disadvantages and advantages that put family variance *above* 80 percent, even values so great that genetic influence would be entirely absent. That genes have no direct effect on our brains is not possible, unless we are the result of divine creation rather than evolution. How to explain the surplus? I will set the limit on home environment at 80 percent of variance explained, and attribute disadvantages/advantages above that limit to the "fairness factor."

Let us imagine that parents try to be fair in the sense of giving all of their children much the same preschool environment, no matter what their genetic potential. Whatever that level of quality may be will obviously privilege the less gifted child more than the more gifted child. How can it not, when it affords the same environment for both, and one is less gifted than the other? Which means that less gifted children get an *extra* advantage from what goes on *within* their home – which will add on to the advantage they get *between* homes. Recall that all those above the median in cognitive performance (a genetic elite) already suffer from a worse match between their genes and environmental quality: for example, if they are slightly above the 84th percentile of genetic quality, they on average come from homes at the 61st percentile; and those slightly below 16th percentile in genetic quality (a genetically substandard group) come from homes at the 39th percentile.

In other words, thanks to this extra advantage, low performers are even more privileged by environment than we had imagined – but only in their preschool years. School soon swamps the early environment and abolishes its extra effects. But while it is pervasive, it will put the less gifted children even closer to their counterparts at the target age (where of course the effects of the preschool environment have long faded away). Similarly, preschool children who are high performers get both their usual disadvantage of the negative gap between their level of performance and the lower cognitive value of their homes, plus the *extra* disadvantage from what goes on *within* their homes, something that expands the score gap between them and their counterparts at the target age.

All of this assumes that siblings differ in terms of genetic potential for cognition. Jensen (1970) notes that more than 17 percent of siblings raised together will differ by more than 24 IQ points. They differ on average by 12 points (excluding measurement error). When parents try to be fair to their young children and, as much as possible, expose them to the same quality of environment, they are not totally successful even in the preschool years. The effects would tend to erode as the infant ages: you try to read as much as possible to both of your children but at some age, one begins to respond with "another book" and the other with "no more book." But you try. It is not a matter that parents supply a quality of environment that averages the difference between their genes. But whatever quality you try to supply to both, it will of necessity provide a more favorable match with the genes of the less gifted child than with the genes of the more gifted. And the overall trend must confer a (comparatively) disadvantageous match for those above the average IQ and an advantageous match for those below.

Therefore, I have calculated the environmental disadvantages/advantages that would explain 80 percent IQ variance from being born into this family rather than that family (*between* family differences). And I hypothesize that any surplus over 80 percent must result from fairness *within* the family. This surplus will vary from one cognitive ability to another, of course, since cognitive abilities differ in terms of how potent parents are likely to be (they can influence Vocabulary more than Block Design).

Given our formula for computing variance, what would a limit of 80 percent entail for the influence of between-family environment? All computations are detailed at the bottom of Table 13. At +2 SD, it entails a maximum of +20.30 points as the gap between an early age and the target age; at +1 SD, a maximum of +9.80 points; at −1 SD, −9.80 points; at −2 SD, −20.30 points. All surplus points beyond these limits are a measure of the fairness effect.

Table 13 summarizes the most striking results. To interpret the table, look at the column of values under "Fairness within family." For those at +2 SD or +1 SD, it tells you how much more disadvantaged they are than seems possible. For those at −1 SD or −2 SD, it tells you how much more advantaged they are than seems possible.

Note how the within-family fairness factor almost always lessens from age 3 to 4 and almost always disappears by the age when children are at school. What a galaxy of results attesting to the presence of parental fairness toward all of their children. For brevity, Table 13 includes only data sets since 1985, and excludes levels and subtests for which the surplus is less dramatic or absent. The full table is in Appendix XIV: the data sets used to derive the values therein are spelled in the various subtest appendices.

Note that subtests that measure parental behavior the child can imitate are amply represented, while subtests that are "test-specific" (measure a skill that has no behavioral analogue among parents, like Block Design and Picture Completion) are absent. Note the high values for Arithmetic, testifying that what children know about numbers before school is primarily determined by what parents teach them.

Table 13 Preschool children: evidence for the fairness factor within families

	Gap early vs. target age	Between-family limit	Fairness within family	Age
Stanford-Binet Vocabulary: average 1985 and 2001				
+1 SD	+11.61	9.80	**1.81**	3
	+11.32	9.80	**1.52**	4
	+6.92	9.80	**NIL**	6.75
−1 SD	−15.14	9.80	**5.34**	3
	−13.78	9.80	**3.98**	4
	−10.48	9.80	**0.68**	6.75
−2 SD	−28.34	20.30	**8.04**	3
	−25.61	20.30	**5.31**	4
	−18.67	20.30	**Nil**	6.75/7
Wechsler Arithmetic: 1992 only				
+2 SD	+28.81	20.30	**8.51**	3
	+29.52	20.30	**9.22**	4
	+19.37	20.30	**NIL**	7
+1 SD	+12.57	9.80	**2.77**	3
	+13.57	9.80	**3.77**	4
	+8.12	9.80	**Nil**	7
−1 SD	−12.33	9.80	**3.53**	3
	−12.33	9.80	**3.53**	4
	−6.88	9.80	**NIL**	7
−2 SD	−24.15	20.30	**3.85**	3
	−19.15	20.30	**NIL**	4
	−6.88	20.30	**NIL**	7
Stanford-Binet Absurdities (=Information): 1985 only				
+2 SD	+30.61	20.30	**10.31**	4
	+29.27	20.30	**8.97**	6.75
+1 SD	+15.48	9.80	**5.68**	4
	+15.48	9.80	**5.68**	6.75
−1 SD	−15.29	9.80	**5.49**	4
	−10.03	9.80	**0.23**	6.75
−2 SD	−29.86	20.30	**9.56**	4
	−16.74	20.30	**NIL**	6.75

(continued)

Table 13 Preschool children: evidence for the fairness factor within families (*continued*)

	Gap early vs. target age	Between-family limit	Fairness within family	Age
	Stanford-Binet Comprehension: 1985 only			
+2 SD	+34.48	20.30	**14.18**	3
	+31.57	20.30	**11.27**	4
	+22.74	20.30	**2.44**	6.75
+1 SD	+25.70	9.80	**15.90**	3
	+24.03	9.80	**14.23**	4
	+10.21	9.80	**0.41**	6.75
−1 SD	−10.96	9.80	**1.16**	3
	−10.76	9.80	**0.96**	4
	−6.68	9.80	**NIL**	6.75
−2 SD	−31.83	20.30	**11.53**	3
	−29.01	20.30	**8.71**	4
	−17.83	20.30	**NIL**	6.75

Calculations – derivation of the maximum between-family limits assumed above:

(1) If family contributed 100 percent of variance, those 2 SD above the mean should be 30 IQ points short of adults at the target age.

(2) However, at +2 SD, it is assumed that no performers come from the bottom 30 percent of the curve of family quality. This reduces the highest possible difference between early age and target age performers to 22.55 points (30.00 − 7.45).

(3) **20.30** divided by 22.55 = 0.90 as correlation, which equals 81 percent of variance explained – close enough to our maximum of 80 percent from between-family effects.

(4) **9.80** points becomes the limit at +1 SD: 15 − 4.11 = 10.89 × 0.9 = 9.80.

(5) The values for −1 and −2 SD are, of course, **−9.80** and **−20.30**.

Dividends of the Age-Table Method

Before we move on to Part II, I will list the theoretical questions the new method clarifies: infant brain physiology is unlikely to reveal who has the best genetic promise for intelligence, why parents influence certain cognitive skills more than others, whether the posited evidential conflict between kinship studies and adoption studies can be resolved, and whether we need a more complex classification of the factors that partition IQ variance. The classes are: genes and environment correlated with genes, the family environment whose influence is eventually eroded, and the kind of environment that is impervious to correlation with genes. The last takes on greater significance when it is divided into chance environment and autonomous environment.

Answers

(1) Parents largely ignore the genetic difference for cognitive ability between their children when trying to develop their preschool competence.

(2) They are most successful for those abilities molded by parent-child interactive behavior, rather than abilities for which children find their own way even in infancy.

Part II

Intelligence

7   The Raven's revolution

Questions

(1) Why was Raven's once thought to be the purest measure of a kind of intelligence largely stable over time?

(2) Now that we know that cognitive progress over time can be rapid, does this mean that Raven's lacks significance?

The Age-Table Method of measuring family effects can also be used on Raven's Progressive Matrices. In 1935, when John C. Raven (then 35 years old) sat at his kitchen table and began to experiment with snowflake designs, he was merely a M.Sc. candidate at the University of London. And yet, what he invented became the most important test in the history of intelligence. It had as its companion the Mill Hill Vocabulary Scale. Spearman, the great inventor of factor analysis, had distinguished two main components of intelligence: *eduction*, or the ability to make sense out of complexity – that is, perceive logical sequences despite distracters that can become very confusing indeed; and *reproduction*, or the ability to store and retrieve what is learned – for example, vocabulary. The Progressive Matrices were designed to measure the former; the Vocabulary Scale, the latter.

Particularly after twin studies began to partition individual IQ differences into variance explained by unequal genetic endowment and by unequal environments, Jensen and his followers deemed Raven's the best measure of intelligence existent; and described what it measured to be an attribute deeply embedded in the human brain and therefore largely impervious to any kind of sudden change.

The presumed stability of intelligence

Intelligence was supposed to be stable precisely because it was genetically influenced. Therefore, it was unlike learning, which is at the mercy of what kind of education your environment provides. Just as, under normal conditions, your genes determine your height relative to others, so your genes determine the neural framework of your brain and its capacity to solve complex cognitive problems. As for the average intelligence of the human race, it had evolved by natural selection and while that process will of course continue, it will be slow.

There is a corollary: some of the behaviors peculiar to the last two centuries can accelerate the rate of change but only because they impinge on genes. First, the invention of contraception and modern medicine is reversing the tendency of the successful to leave behind more offspring than the unsuccessful; the former can now limit their number of children and the latter can live long enough to breed prolifically. Second, human mobility is lessening the number of isolated people that inbreed to the detriment of their offspring. But even these "speed ups" should not have a great impact over a few generations, particularly since they offset one another. One degrades the gene pool, while the other maximizes its potential because it diminishes the chance that two deleterious recessive genes will be paired during sexual reproduction.

Why was Raven's deemed the best measure of intelligence? It best predicts how people perform over a whole spectrum of cognitively demanding tasks, ranging from how to complete a number series, to how to solve three-dimensional jigsaw puzzles, even to size of vocabulary and general information. In addition, it seemed to be a test whose content came closest to being cultur-ally reduced and would therefore be familiar to all peoples. It used simple things like circles and triangle and squares.

Jensen (1980) believed it could give meaningful results for Polar Eskimos and Kalahari Bushmen as well as for Americans of all races. If Martians landed on earth, it would tell us whether they were brighter than we were. He made it a virtue that Raven's does not predict achievement as well as Wechsler tests. Intelligence + motivation = achievement. Wechsler tests measure both intel-ligence and motivation (they include subtests like whether you were motivated to read and collect a lot of general information). However, Raven's measured intelligence alone: it was too "factor pure" a measure to predict achievement any better than intelli-gence alone could predict achievement.

Jensen's firm conviction that intelligence was stable and Raven's its measure was conveyed in his response to Flynn (1984), which showed that IQ scores on Wechsler tests had increased by almost a full SD over only 46 years, at least in America. He allowed that since Wechsler IQ measured school skills to some degree, it might escalate; but he predicted that performance would not increase over time on a culturally reduced test like Raven's (Flynn, 1987). He was mistaken: Raven's gains over time were greater than those on any other test in fourteen nations, all those for which data were available. In fact, Raven's turned out to be the least cul-ture-proof of all the IQ tests in existence.

Table 14 updates adult Raven's gains over time from seven nations and nine groups, and compares them to the Vocabulary gains of American adults. Jensen was correct about the effect of changes in schooling between generations as a cause of Wechsler gains. As the percentage of Americans (aged 25 and above) with some tertiary education rose from 12.1 to 52.0 percent, the extra years of schooling gave them a 17-point IQ gain on the WAIS Vocabulary subtest or a rate of 0.318 point per year, which is very nearly the highest of the WAIS subtests (Flynn, 2012a, pp. 100–1).

Table 14 Raven's and Vocabulary: sensitivity to environment over time

Locale	Age	Dates	Gain IQ points	Rate per year
Scotland	77	1921–36*	16.50	1.100
Belgium (Flemish)	18+	1958–67	7.82	0.869
La Plata (Argentina)	19–24	1964–98	27.66	0.814
Belgium (Walloon)	18+	1958–67	6.47	0.719
The Netherlands	18	1952–82	20.10	0.667
Israel (female)	17.5	1976–84	5.09	0.637
Norway	19–20	1954–68	8.80	0.629
Israel (male)	17.5	1971–84	7.35	0.565
Great Britain	18–67	1942–92	27.00	0.540
Raven's Matrices	**Adults**	**Circa 1940–84**	—	**0.727**
Wechsler Vocab	**20–74**	**1953.5–2007**	**17.0**	**0.318**

*The dates refer to dates of birth.

Sources:
Flynn, 2012a, Box 11 and Table AI3 (in that table, the year for the standardization of the WAIS-IV should be 2007 not 2006); Staff et al. (2014).

However, the average Raven's gain, at least during the peak period of IQ gains (1940–84) was 2.3 times as great at 0.727 points per year. Whatever it is that Raven's measures is hardly impervious to environmental change from one generation to another.

The malleability of what Raven's measures

Raven's measures the ability to see logical sequences in a series of images that do not correspond to the concrete objects that exist in the world of sense. The correct answers to every item are dictated by a coherent set of rules that range from simple to complex. The most illuminating analysis of what Raven's measures was published only recently. Using the Advanced Progressive Matrices test, Fox and Mitchum (2013) allow us to analyze what has altered in people's minds when one generation scores higher on Raven's than the last. The following analysis is in my language (reproduced from Flynn, 2012a, pp. 284-6). However, we met at the University of Richmond, and they confirmed that my interpretation is compatible with their analysis.

At the beginning of the twentieth century, people just beginning to enjoy modernity were still focused on the concrete objects of the real world. They wanted to manipulate the real world to their advantage, and therefore the representational images of objects was primary. If you are hunting you do not want to shoot a cow rather than a deer; if a bird is camouflaged in a bush, you flush it out so its shape can be clearly seen. Raven's poses a problem that is quite alien to your "habits of mind": you must divine relations that emerge only if you "take liberties" with the images presented. It is really a matter of perceiving analogies hidden behind distracters. I will present a series of analogies (the first three are my own) to illustrate the point.

(1) Dogs are to domestic cats as wolves are to (wild cats). Presented with these representational images people a century ago would have no difficulty.

(2) ■ is to ♦ as ↑ is to (→) where the choices are ↑, →, ↖, and ↗. Here you must ignore everything about an image except its shape and position. Just as the square has been rotated a half turn, so has the arrow.

(3) □ is to / as O is to (|) where the choices are ∅, ⊖, |, and ⊗. Here you must ignore everything but the number of dimensions: the analogy compares two-dimensional shapes to one-dimensional shapes and all else is irrelevant. Representational images are of course three-dimensional, so such a contrast requires being well removed from them.

(4) &#B is to B&# as T&T is to ##(enter what symbol fits). This is an item from Fox and Mitchum (2013) that illustrates the kind of analogical thinking you must do on the Advanced Raven's Progressive Matrices.

Note that the right answer in the fourth item has been left blank. Since no alternatives were presented to choose from, you had to deduce that "&" is the correct answer. I got it right, which was reassuring given that I was then 78 years old, by reasoning as follows. In the first half of the analogy, all that has altered is the sequence of symbols: labeling them 1, 2, 3, they have become 3, 1, 2. Applying that to the second half of the analogy, T&T changes to TT&. Clearly you are supposed to ignore the fact that the doubled letter (TT) has changed to a doubled symbol (##), so the right answer is ##&. This would really discriminate between the generations. We have moved far away from the "habit of mind" of taking pictorial images at face value; indeed, we are interested only in their sequence and treat images as interchangeable if the logic of the sequence demands it.

The key is this: anyone fixated on the literal appearance of the image "T," as a utilitarian mind would tend to be, would simply see no logical pattern. Contrast this with Wechsler Vocabulary. The etiology of enhanced scores over time would be quite different. People over time, thanks to the bonus of more education, simply accumulated a larger store of core vocabulary and got no bonus from the shift from utilitarian toward "scientific" thinking. Excepting of course for words that labeled abstractions (like species), which now appeared in the new subjects taught.

Fox and Mitchum (2013) classify Raven's items in ascending order of "relational abstraction": more specifically, "for analogical mapping when relations between objects are unrelated to objects themselves." Once again, in example number 4, the relationship can be derived only if one sees that a "T" does not have to retain its identity as a "T." Their core assumption was that "analogical mapping of *dissimilar* objects is more difficult than mapping *similar* objects" (italics mine). I certainly found this to be true. The fact "TT&" had to be translated into "##&" rendered the item harder to solve. And if I were my father (born in 1885), and wedded to taking images at face value for reasons of utility, I suspect I would have found it insuperable.

They analyzed the performance of two samples of young adults tested in 1961 and circa 2006 respectively. They found that as the degree of deviation toward the abstract increased, certain items became less predictive of performance within the two generations than between the two generations.

We now know why Raven's scores are so sensitive to environmental change over time. Like our ancestors, we can still use logic to analyze the concrete world. But we have entered a whole new world that allows us to use logic on symbols far removed from the concrete world. We organize the concrete world using abstract concepts that are not represented there.

Pre-modern people see fish as having nothing in common with crows. You can eat one and not the other; one swims, the other flies. We use DNA analysis to divide living creatures into categories that are non-observable but offer understanding, and this language has become that of every person who has been exposed to several years of formal schooling. We know that bacteria differ from one-celled animals, that whales are more akin to land animals than fish, and that the tiny hyrax is more akin to the huge elephant than to the rodents it resembles. We know that stars are different from planets (although they look the same in the sky) and, indeed, our whole picture of the universe (and even our approach to explaining human behavior) is based on logic and abstractions. We are exposed to the symbolism of algebra. No one has ever observed an "x."

In other words, using logic on symbols detached from concrete reality has become a habit of mind in no way alien to us. These skills are not merely useful in mathematics and science and computer programming (programmers do very well on Raven's). They help us to create (and comprehend) a non-representational map of the London underground, or an organizational map that functionally relates the tasks a complex business organization performs. We are more ready to engage with Raven's because the rise of modernity altered our perspective. And the rise of modernity has occurred over only a few generations. Only a test that is sensitive to the new minds that modernity has put into our heads could measure something so malleable. Raven's, more than any other test, *is a barometer of the stages of modernity* and thus continues to play a crucial role in the study of intelligence.

It can tell us how far people have gone down the road that enhances our ability to solve the cognitively complex problems of the modern world. Khaleefa et al. (2009) report WAIS-R gains in the Sudan between 1987 and 2007. The gains by subtest (Flynn, 2012a, pp. 62–3) are largely on those that exposure to the mass media would dictate – that is, immersion in a visual culture, spatial imagery, and a heightened speed of information processing. Their gains were very low on the subtests most responsive to formal schooling – that is, Information, Vocabulary, and Arithmetic – signaling the fact that their school system has been retarded by a traditional "Muslim curriculum," which was imposed in 1990. Schooling was also disrupted by over two decades of civil war. The majority of the South Sudanese are still illiterate. Due to lack of formal schooling and a scientific perspective, I predict that they would score well below advanced nations on Raven's. It would register that their exposure to modernity has been superficial and that they have not really developed new habits of mind.

Why Raven's gains vary with age

The Dickens/Flynn model (Dickens and Flynn, 2001) predicts that the size of the IQ advantage will vary depending on the age at which we compare a later cohort (say, those born in 1936) with an earlier cohort (say, those born in 1921). Both of these groups live their own lives. During those lives the causal factors that differentiate the later from the earlier cohort vary greatly. This means that the IQ gap that separates the two will vary in magnitude with age according to the potency of the differential factors that kick in at each age. This prediction remained only a prediction until a recent study. As Staff et al. (2014) say, their study is the first to compare two cohorts at two different ages.

The Lothian Birth Cohorts were born in 1921 and 1936 respectively. They included almost every child born in Scotland in those years (and still attending school there at the age of 11). Both were tested on Raven's Progressive Matrices: the later cohort outscored the earlier by 3.7 IQ points at age 11 and by 16.5 IQ points at the age of 77. The difference is huge: the rates of gain differ at 0.247 points and 1.100 points per year over a period of fifteen years. If anything, the gain in old age is an underestimate: the earlier cohort lost more people by death (earlier death is negatively correlated with IQ) than the later. The differing gains must reflect the relative potency of the causal factors that separated the cohorts at those two ages. What might these be?

When you test two cohorts at the age of 11, they both have approximately the same number of years of formal schooling and this serves as a leveler: the small IQ gap would reflect only the fact that the later cohort came from homes a bit higher in SES and any progress made in the quality of schooling. I hypothesize that the gap would double at the age of 21: thanks to more students going on to tertiary education, the later cohort would have more years of formal education.

This speculation is based on Table 14 and its rates of gain for the seven cases in which subjects were aged 17.5 to 24. They average at 0.700 points per year, which contrasts with the average of 0.326 points per year for four samples aged 7.5 to 16 from four developed nations (Flynn, 2012a, Box 11: I have omitted the outlier from Leipzig). By age 35, the influence of schooling would have faded in favor of the later cohort working at more cognitively demanding jobs. No data reveals whether this would confer a greater or lesser advantage than was present in the university years. At the age of 70, I would anticipate a lessening of the gap, since both cohorts would have retired from work – *except* that the later cohort would be far more healthy and alert due to modern medicine alleviating the illnesses of old age, more exercise and better diet by older people today (I still run at 81 while my father took no exercise after age 14), and the greater cognitive demands of leisure activities. Once subjects reach the age of 77, we have real data. We know that the three factors named produce a huge gap (16.50 points for two cohorts only ten years apart) unlikely to be matched at any earlier age.

I have often rejected the hypothesis that generational IQ gains reflect gains in heath and nutrition, at least in advanced nations since 1950. This was because we were looking for them in the wrong place: we thought they would weigh in at the beginning of life (they do not); rather, they weigh in at the end. At any rate, we now know that Raven's is not merely sensitive to the global environment enriched by modernity. It is also sensitive to each and every one of the particular factors that have triggered IQ gains over time.

Raven's and family effects

Our new method can reveal whether or not family differences affect performance on Raven's. If Raven's is sensitive to most environmental factors, it would be odd if it was impervious to the cognitive quality of family environment. Before the influence of the home wanes, how much of Raven's variance (setting chance aside) is accounted for by family effects uncorrelated with genes?

Three versions of Raven's are matched to different ages. The Coloured Progressive Matrices (CPM) are for preschool children and school children up to age 11; they were normed in 1949, 1982, and 2007. The Standard Progressive Matrices (SPM) overlap with the CPM and are for schoolchildren through age 15.5. However, they were also used for adults until the development of the Advanced Progressive Matrices (APM). They were normed on schoolchildren in 1938, 1979, and 2008 (when they became the SPM plus) and on adults in 1942. The APM are the current test for adults, normed only in 1992.

The age overlaps between these tests (and a table that equates SPM scores with APM scores) provide raw scores all the way from preschoolers to our adult target age (the age after which performance begins to decline). The years when the tests were normed suggests three sequences of data: early data linking 1949 with 1938 with 1942 with an adult target age of 22.5 (20–25); intermediate data linking 1982 with 1979 with 1992 with a target age of 25 (18–32); and the latest data linking 2007 with 2009 with 1992 also with a target age of 25. Note that the 1992 norming of the APM (and their table) has to do double duty as the culmination of both the intermediate and latest sequences. The data and the calculations are in the Raven's Appendix (Appendix III). The method for determining the magnitude and persistence of family effects on Raven's is of course identical to that used on all of the Wechsler and Stanford-Binet subtests (see Appendix III).

Table 15 compares the most recent Raven's results to the most recent results for the Wechsler Vocabulary subtest, the subtest that showed the most persistent family effects of all. The table shows how many IQ points the typical performer at various ages would forfeit because of family disadvantage above the median, and how many IQ points they would gain because of family advantage below the median.

From ages 4 to 9.5, the values for Raven's and Vocabulary are almost identical. At ages 12 and 15, the Vocabulary values are larger above the median and the Raven's values larger below the median. Unfortunately, age 15.5 is the oldest age that the Raven's data allows to be normed on the target age. But note this: the peak age of Raven's performance is 25 and that of Wechsler Vocabulary is about 50. It is possible that family effects on Raven's persist even after performance begins to decline with age, but by definition our method cannot detect them. If these effects exist, they might not only match Vocabulary at ages 17.5 to 20–24 but also boost all Raven's values at earlier ages (all effects are additive from older ages down to younger ages).

Table 15 Comparison of the most recent Raven's and Wechsler Vocabulary results

Raven's: all ages normed on the target ages of 18–32 (25)							
Results from the 2007/2008/1992 standardizations							
Percentile 4.25	7.50	9.50	12.50	15.50	17.5	18	20–24
95 +19.90	+11.62	+5.25	−2.45	−0.975	—	—	—
82.5 +17.88	+10.82	+7.05	+1.98	+2.33	—	—	—
17.5 −16.62	−8.51	−4.57	−7.04	−4.75	—	—	—
5 −19.53	−11.43	−7.49	−8.38	−7.67	—	—	—
Wechsler Vocabulary: all ages normed on target ages of 45–54							
Results from the 2002/2002/2007 standardizations							
Percentile 4.00	6.75	9.25	11.50	14.50	17.5	18	20–24
98 +19.72	+8.91	+4.45	+1.91	+0.75	+0.25	−2.25	−1.25
84 +13.77	+9.52	+7.48	+5.77	+4.42	+5.25	+4.75	+3.75
16 −13.18	−8.77	−6.23	−5.02	−1.68	−4.18	−4.11	−3.96
2 −26.72	−15.25	−8.96	−7.75	−4.42	−5.25	−5.00	−4.50
Cor. 1.134	0.688	0.463	0.356	0.197	0.278	0.234	0.213
% var. 128.28	47.30	21.45	12.67	3.89	7.71	5.47	4.54

Table 16 compares results for all three of the Raven's sequences – that is, the most recent, the intermediate years, and the early years. The intermediate and early results also match Vocabulary at ages 9.5 and above, but they differ from the most recent data by showing no additional family effects at age 8 and below. They do show large family effects at those ages, of course, but still, in almost all data sets the family is even more influential at young ages, before it has much competition from schools and peers. The early results differ from the other two at age 12 (and above) by reversing the results above and below the median: they differ by showing high values above and low values below.

There is good reason to place more trust in the data as we approach the present. The standardizations seem more rigorous (Raven, 2000; Raven et al. 2008a). But more important, gains over time meant that high scores on the SPM began to be depressed by ceiling effects: too many people were approaching a perfect score

Table 16 Family effects by age on Raven's at three times

The first and second sets of standardizations are normed on the target age of 18–32 (25); the third is normed on the target age of 20–25 (22.5).

Percentile	+/– SD	4.25	5.50	7.50	8.00	9.50	12.50	14.00	15.50
Results from the 2007/2008/1992 standardizations									
95	+1.645	+19.90		+11.62		+5.25	−2.45		−0.975
90	+1.282	+20.10		+11.43		+7.28	+1.91		+1.80
75	+0.645	+15.66		+10.21		+6.82	+2.04		+2.86
25	−0.645	−13.70		−6.09		−2.15	−6.30		−2.33
10	−1.282	−19.53		−10.92		−6.98	−7.77		−7.16
5	−1.645	−19.53		−11.43		−7.49	−8.38		−7.67
Results from the 1982/1979/1992 standardizations									
95	+1.645		+10.53	+10.99		+3.76	−0.425		−0.975
90	+1.282		+11.45	+8.45		+3.61	+0.81		+1.80
75	+0.645		+4.20	+4.60		+2.71	+2.86		+2.86
25	−0.645		−9.10	−10.29		−2.15	−8.53		−2.33
10	−1.282		−10.25	−10.63		−8.29	−8.87		−7.16
5	−1.645		−10.25	−10.63		−8.80	−9.38		−7.67
Results from the 1949/1938/1942 standardizations									
95	+1.645			+13.53	+13.73	+13.82	+11.30	+7.78	
90	+1.282			+9.06	+6.60	+7.59	+4.74	+4.41	
75	+0.645			+12.06	+9.43	+6.63	+2.79	+2.24	
25	−0.645			−3.87	−7.54	−6.55	−3.96	−2.24	
10	−1.282			−14.16	−14.16	−5.52	−2.93	0.00	
5	−1.645			−14.16	−14.16	−5.52	−2.93	0.00	

of 60, and this limited those at the 90th percentile from opening up much of a gap over those at the 75th percentile, and those at the 95th percentile had virtually no advantage over those at the 90th percentile. The presence of ceiling effects for adults was cured only by the addition of the much harder APM in 1992. And the ceiling effects that even older schoolchildren encountered on the SPM were cured only when the test was toughened in 2008. The new SPM plus added some difficult items that replaced easier items in the old SPM.

The contemporary importance of Raven's

Contrary to its reputation, Raven's is sensitive to every environmental influence going: the progress of modernity, factors that trigger IQ gains over time such as schooling and health, and family effects. At least, it maximizes the persistence of family effects as much as Vocabulary and that means as much as any other test. It has lost its role as the purest measure of a kind of intelligence that alters only at the slow pace at which the genetic quality of humanity alters. But whenever we study the cognitive abilities of humankind, Raven's emerges as an indispensable instrument of measurement. Its environmental sensitivity is its salvation.

The fact that Raven's skills and Vocabulary skills show exactly the same pattern of development counts against a hypothesis we will encounter later. There is no reason to posit that Raven's measures an underlying factor (fluid g) that is "invested" in acquiring skills like Vocabulary. You could as easily argue the reverse. The two seem to develop together by way of reciprocal causality. Better verbalization helps you to analyze better (solve Raven's problems), and better analytic skills put you with cohorts who are above average in Vocabulary, and so forth.

Answers

(1) The traditional role of Raven's was based on premises that have been falsified: that there was a sort of intelligence largely impervious to culture, and that Raven's was a culturally reduced test.

(2) However, the very fact that Raven's is sensitive to cultural evolution in the form of the rise of modernity makes it the best barometer of cognitive progress over time.

---

8   Learning from astronomy

Questions

(1) Why does astronomy need a meta-theory – which is to say, need heuristics?
(2) How do the heuristics of astronomy differ from its theory-embedded concepts?

Thus far, most of this book has been a contribution to the science of individual differences in intelligence. Previously, I have contributed to the science of how cognitive abilities alter over time. I will now venture into the theory of intelligence, partially for its own sake and partially to put my contributions into context.

I want to show that every science needs something I call a meta-theory just as much as it needs fertile scientific theories and, indeed, that there is a relationship between the two. The latter explain and predict phenomena, whether they pertain to the motions of the planets or to intelligent human behavior. The former consists of one or more heuristics – which is to say, concepts that offer advice to theory builders. This advice can be good or bad in the sense that it can set strictures on scientific theories that limit their explanatory potential. The quality of the meta-theory should be measured in terms of the quality of the scientific theories it engenders. In addition, there must be a body of data that allows both verification and falsification. This last is not passive. Some times new data emerge that signal either a flaw in an existing theory and, therefore, the need for a new theory, or the need for a new meta-theory because it appears that current heuristics are giving bad advice about theory building.

I will begin by demonstrating the need for meta-theory even in the most rigorous of the sciences: astronomy, as inclusive of physics and cosmology. When we review the history of astronomy, we will find that the failure to realize that every science has a meta-theory encourages three kinds of error:

(1) There is a failure to face the fact that new evidence shows that a hitherto successful theory needs to be revised (type-one error). Often this occurs simply because scientists, like anyone else, are reluctant to embrace new ideas. But sometimes, it is because you are under the spell of a heuristic that makes new theories seem "impossible."

(2) When that occurs, the only solution is to take the more radical step of revising the heuristic itself, and failure to do that (type-two error) will halt the progress of science. Clearly, if you are not aware that every science has a meta-theory or heuristic, you will be inhibited from seeing the need to revise it. You will have been actually operating in accord with some heuristic's piece of advice, of course, but doing so semi-unconsciously.

(3) The most serious kind of error is when you have confused the role of meta-theory and theories. You take what has been a fruitful scientific concept (say g) and elevate it to the role of heuristic or advice giver (type-three error). For example, you may say that any plausible theory of IQ gains over time must show that these gains are on the g factor. This error in a sense combines the two previous kinds of error. It is particularly tempting because scientists are

accustomed to valuing precision. However, it is the very precision of a theory-embedded concept that makes it inappropriate for it to be more than that.

We do want precision on the theory level, but on the meta-theory level we want breadth. Heuristics or advice to theory builders should point them in a certain direction, of course, but it should be broad enough to allow a number of conceptually inconsistent theories to compete with one another to see which one explains the evidence best. As we shall see, some astronomers elevated a concept (a mechanical model of the universe) from its proper sphere (as a theory-embedded concept) to the level of a meta-theory heuristic. They simply used it to rule out any non-mechanical theory of the universe as non-viable. This is the best method of strangling competing theories. To be viable, they cannot be fresh and new; they must be variations on an established theory. Once your theory has become a piece of advice to theory builders, any evidence that counts against it looks like bogus evidence.

Greek astronomy and its concepts

The Greeks invented the science of astronomy. They had an interesting heuristic that gave advice to theory builders, namely: *the observed motions of the heavenly bodies should be reduced to circular patterns.* The veneration of the circle had roots that are unclear, but circular motion was thought to be inherently perfect. They believed that the stars rotated around the center of the universe on a spherical globe. A straight line was the natural path for an object moving on a plane, but they knew that a circle was not necessarily the natural path for motion on a curved surface. For example, they knew that a plane that intersected a cone could produce a curve we call an ellipse. Perhaps they felt as they did because the heavenly bodies were considered divine and circles were considered to be the most beautiful curve (appropriate to the motions of a divine body). In any event, the fascination with the circle endured right up through the time of Kepler. One reason it endured so long is that the heuristic of the Greek astronomers did not dictate that they produce a workable mechanical model of the universe. That came later.

Their greatest astronomer, Ptolemy, was aware that the heuristic was broad enough to allow *competing theories to be tested by evidence* and insisted on such. Some theories embedded the concept that the earth was the center of the universe; others that the sun was the center of the universe; later, Tycho Brahe made Mercury and Venus moons of the sun but still believed that the sun circled the earth. Ptolemy appealed to evidence: if the earth rotated, everything not attached to it would be thrown back and fall to the west: convincing if one was unaware of the law of inertia.

As time went on, evidence posed another seemingly insuperable objection to the earth's motion around the sun – namely, the absence of a stellar parallax. If the earth moved in a circle, and the stars like the sun were fixed, then at the extreme points of the earth's orbit a star should appear to be on a different angle, say to the west at one point, to the east at another. But until quite recently, no observation showed a stellar parallax. This is because the stars are so very far away. No one could conceive of this: they did not have the concepts (like a light-year) that we use as a metric of the immensity of the universe.

Ptolemy's theory worked brilliantly. By using epicycles – that is, circles that rotated on other circles rather like a Ferris wheel – he accounted for all the observations of the heavens available in his day. From the time of the Babylonians, a large number of observations had been accumulated that furnished a body of data that could be used to verify or falsify. Prior to the Babylonians, astronomical theory was not very advanced. The Egyptians believed the sun disappears at night because it is eaten by a sky god (whom they identified with a cow) and is reborn the next morning. Only postmodernists still think that this theory is plausible: they believe accounts of the heavens are texts subject to an infinite number of interpretations.

From Ptolemy to Kepler

Even before the telescope was invented, thanks particularly to Tycho Brahe, the body of observations available to astronomers became more exact and the suspicion arose that the number of epicycles needed to account for the observations was going to be ridiculous. However, despite the new evidence, most astronomers continued to commit a type-one error: they simply could not believe that a theory as fruitful as that of Ptolemy must be abandoned. This error was excusable because the astronomers were wedded to Ptolemy unless they did something far more difficult. What they really needed was to conceive of a whole new heuristic: they needed to abandon the advice that all theories must utilize circular motion in favor of new advice. In effect, they were also committing a type-two error.

Kepler was less addicted to Ptolemy's theory because he was a sun-worshiper who thought that the sun must be the center of the universe. Unfortunately, so long as everyone stuck to the old heuristic (circular motion), the evidence was equivocal. Galileo stuck to the old heuristic and there were three things against his model. First, the planets actually move around the sun in ellipses, a sort of squashed circle. In addition, the sun is not really at the "middle" of the planetary orbits; rather, it is at one of the two foci of the ellipses. Therefore, Galileo still had to posit epicycles and in order to minimize their number, he had to put the "center" of the solar system at a point in space near the sun! Second, even when he did this, he needed more epicycles (circles on circles) to cover the observations than Ptolemy did. Finally, there was still the "irrefutable" lack of a stellar parallax.

The movement of the sun contradicted a literal interpretation of scripture ("the sun runs his course like a strong man in the heavens"). The Church was willing to alter a literal interpretation if science had proved something to the contrary. Cardinal Bellarmine knew a lot of astronomy and knew that the new theory needed more evidence. He told Galileo he could pursue it as a hypothesis but could not assert that it was true (so long as it was manifestly suspect). Galileo was arrogant enough to imply that only idiots doubted its truth and painted the pope as an idiot. He had to recant (he is said to have been a "broken man" – that did not prevent him from developing the science of terrestrial dynamics in his spare time).

Initially, Kepler was under the spell of the old heuristic (type-two error). The orbit of Mars deviates most from the circular of the planets known in his day. In his magnum opus, he remarks that he could make sense of Mars if only it were possible for planetary orbits to be elliptical. Finally, a few hundred pages later he faces the awful truth: its orbit really is a big, squashed, ugly-looking ellipse. He consoled himself in a later book with the notion that you could fit the five perfect solids within the planetary orbits so long as you filled gaps with the musical harmonies. Faced with what the observations implied, he had abandoned the old heuristic. But he did not supply a new one that offered new advice to theory builders. Kepler had to offer his laws about the solar system as brute descriptions with no theory that linked them.

Newton and his concepts

Fortunately, with knowledge of magnetism, a new heuristic began to dawn: *that the mass and location of heavenly bodies should guide theory builders*. After all, magnetism showed that two bodies could influence one another at a distance. And the sun was so big, and the earth so small, and the earth was so much nearer the sun than any other huge body, it looked like location ought to count. This new heuristic, like all good heuristics, was broad enough to allow for the existence of a variety of theories with different theory-embedded concepts that could be tested against the evidence.

Descartes and Newton stepped forward with competing theory-embedded concepts both of which fell under the new heuristic. Descartes posited (without evidence) that the sun rotated on its axis and created a whirlpool in the ether that carried the planets around in their orbits. Newton's theory embodied the concept of universal gravitation: all heavenly bodies attract one another in proportion to their mass and inversely as the distance (between them) squared. A whole chapter of his *Principia* is devoted to the mathematics of a whirlpool, demonstrating its inadequacy; this seemed irrelevant but he had Descartes in mind. The mathematics of his own theory dictated the existence of not only Kepler's three laws of planetary motion but also the laws (discovered by Galileo) that govern the movements of objects within the earth's gravitational sphere.

The absence of the stellar parallax was still unsolved: it was only in modern times that we finally realized that the stars were very far away and developed the instruments to measure the small change of angle that exists thanks to the earth's motion. People at that time just decided to set the stellar parallax aside because the new theory was so successful. Indeed, most nineteenth-century physicists eventually made a type-three error: they elevated Newton's key concept to the level of a heuristic. Its role became that of giving advice to all possible theories. They said that they would countenance no theory of the universe unless its model could be built in a machine shop. Thus they ruled out anything that deviated from the detail of Newton's mechanical universe: namely, objects located at a point in absolute space organized in three dimensions.

They began to be irritated by one observation that did not fit: Newton's equations predicted the motions of all planets save Mercury, whose orbit was deviant.

Given their mind-set the only possible explanations were observational error (quickly dismissed) or an undiscovered planet closer to the sun than Mercury that pulled Mercury out of position. The academy of Dijon gave a prize to the discoverer of this planet and it was named Vulcan (it was a sunspot). If only they had waited. It can be shown that Mercury's orbit is correct if you make the (totally gratuitous) assumption that the sun's gravitational pull suddenly decided to move from its center to its surface in the case of Mercury (and only Mercury). Einstein had a better idea. By a flight of genius, he replaced Newton's theory in its entirety. It was reduced to a special case within Einstein's equations, which is why we can still use it over "short" distances, rather like treating the earth as flat when we lay out a tennis court.

The theory shift from Newton to Einstein

Einstein accepted *the same heuristic* that guided Descartes and Newton: the concept that the mass and location of heavenly bodies should guide theory. However, he abolished the concept of gravity: bodies pulling on one another across space even though they did not touch (which had always seemed odd: much less plausible than people sending thoughts across space from one mind to another).

What the mass of a heavenly body did was warp a space-time continuum in its immediate vicinity. Imagine heavy balls (the stars) dropped onto a floating blanket. Each star including the sun would create a funnel in its vicinity and would lie at its bottom. Planets in motion near the sun would spin around in the funnel-shaped space in an ellipse; and the planet closest to the sun (Mercury) would spin in the space with the maximum warp and be slightly deviant. Einstein used the geometry of Riemann applied to four dimensions (three space, one time) rather than the three-dimensional geometry of Euclid. The integration of space and time was itself illuminating: time passes more slowly for objects that move through space at very high speeds. His theory suggested many other fruitful hypotheses about the maximum speed and trajectory of light, the vibration of atoms on the sun, and so forth.

Einstein's four-dimensional model did not unite gravity and electromagnetism. Also, sub-atomic phenomena upset him. The quantum theory that explained these involved probability, which he found obnoxious. He spent the last years of his life in a futile effort to accommodate sub-atomic phenomena within his theory. Physics divided into two areas, super-atomic and sub-atomic, each having its own meta-theory or quasi-heuristic. Take the mass and location of heavenly bodies into account on the one hand, and try to understand elementary particles on the other. The additional heuristic was broad enough to allow many scientific theories to compete: they could vary in terms of the kinds of their elementary particles (and how many there were), how many dimensions the space they operated in, and how many universes they populated (one up to infinity).

Hopes for unification

The failure of Einstein has not deterred cosmologists from seeking unification. They dream of a theory that will unify the four forces of nature: gravity, electromagnetism, and both the weak and strong forces within the atom. They want to account for new actors in the gravitational equation, dark matter and dark energy. Unlike psychology, I can see no obvious reason why they must fail. We may have found the "Higgs boson" and the search for the "monopole" continues (there may be only one in the universe, so that search may take some time). At any rate, we leave them here. Let us see if what we have learned from the history of astronomy will help us understand the history of the science of intelligence.

Answers

(1) Astronomy has benefitted from changes in meta-theory: new heuristics that gave new advice to astronomers and physicists, but were not so narrow as to outlaw competing theories.

(2) On the "lower" level of scientific theories, the embedded concepts of astronomy (gravity) were narrow and that was why its scientific theories could generate predictions. Except for the Newtonians in the late nineteenth century, theory-embedded concepts never rose to try to play the role of a heuristic; and when they did they were an impediment to scientific progress.

9 The meta-theory of intelligence

Questions

(1) What role do heuristics or quasi-heuristics play in the study of intelligence?
(2) How do heuristics leave room for evidence – that is, permit a variety of scientific theories free to compete with one another?

About 100 years ago, beginning with the testing of army recruits in 1917 in America, psychologists began to collect data that could be used to test theories of intelligence. They used the Stanford-Binet of 1916 and other tests that were the forerunners of Wechsler's IQ tests. These tests are an implicit response to a heuristic that has given advice to theory builders from that time to the present.

The heuristic of intelligence

I will summarize the content of that heuristic (for a full statement, see Flynn, 2009, pp. 53–4). It emphasizes the following traits.

> **Mental acuity**: the ability to solve on-the-spot problems we have never encountered before. The problems deemed important vary, of course, from society to society.
>
> **Habits of mind**: for example, the extent to which people are accustomed to use logic to analyze problems, and

to deal with symbols increasingly detached from the concrete world.

Vocabulary, knowledge (including mathematics in our culture), **and information**: the more you have, the wider the range of problems you can attack.

Speed of information processing: the quicker one can assimilate data the better, particularly if problems must be solved within a time limit.

Rote memory and working memory: the larger the amount of relevant information (and conceptual rules) you can keep in mind, the better equipped you are to arrive at a solution.

Note that the whole content of this heuristic is cross-cultural in the sense that other societies prioritize problems unlike ourselves: Australian aborigines put mapping skills (to find water) well above the skill of using logic on abstractions so useful in formal education.

Wechsler thought within the context of modern Western society. He did not construct a theory of intelligence. He made fun of the Stanford-Binet for the cutting lines it used to separate levels of mental ability: gifted, high normal, low normal, mentally retarded. He noted that the scores all ended exactly in zero and said that this was astronomically improbable. All of Wechsler's new cutting lines ended in exact multiples of standard deviations, which was no less improbable.

Instead of theorizing he jumped straight from the heuristic to the measurement of intelligence. The body of quantifiable data we have, particularly its quality, is based on how perceptively he made this leap. This can be appreciated by listing his subtests: Block Design and Object Assembly and the pictorial tests (on-the-spot problem-solving), Similarities (classifying based on abstractions), Vocabulary, Information, Comprehension (of the concrete world that surrounds us), Arithmetic (mental arithmetic also tests for working memory), Coding (speed of information processing), and Digit Span (forward digit span for rote memory, backward for a crude test of working memory). The data his subtests yield relate not only to testing individual differences in cognitive ability at a given time (within advanced societies), but also to tracing trends in cognitive ability over time, and offer the potential to identify what brain processes correlate with various kinds of cognition.

The positive manifold and twins

After I had charted massive IQ gains in America using both the Stanford-Binet and Wechsler's tests, I got interested in the fertility of scientific theories – namely, theories with embedded concepts that suggest fruitful and testable predictions. Immediately I became aware of the basis of Jensen's theory, his emphasis on the "positive manifold": the fact that those who do better on one Wechsler subtest tend to do better on all of them. My first opinion was that this was easily explained and not very promising. Notes I took back in 1984 record my pristine (unlearned) reaction. Off the top of my head, I assumed that four things were enough to explain the positive manifold.

> PPC: Everything we do, including cognition, presupposes physiological factors, so I coined the term the physiological prerequisites of cognition (PPC). Undoubtedly there was an optimal brain behind good IQ test performance and it probably had features that enhanced all cognitive tasks, more neurons, better connections between neurons (I was ignorant of the role dopamine sprayers play), and an optimal blood supply that nourishes all parts of the brain (look at what hardening of the arteries does to cognition).
>
> ER: I called the physiological factors "prerequisites" rather than "sufficient conditions" because the conscious mind seemed equally important. Freed of coercion (forced to

do your homework), you must rely on your personal autonomy: plan how to use your brain (an executive role or ER). You might of course decide to use it in a way that favored one Wechsler subtest over another (to read and build a strong vocabulary but never do arithmetic).

FR: However, it was highly probable that different kinds of mental exercise were functionally related (FR). The New Zealand field hockey team once won the Olympics by outlasting their competition: when they came home every one of them completed a marathon. Clearly endurance training paid off for more than one sport. If you read widely, you not only build up a big vocabulary but also amass a large fund of general information. On the other hand, it seemed clear that vocabulary exercise had a weaker functional relationship with arithmetic given that so many are good at one and not the other.

CYK: However, there is another factor that would tend to boost both verbal and arithmetical skills to some degree – namely, the company you keep (CYK). In my high school, the best students tended to bind as friends and this peer group enhanced all of our cognitive skills: I learned more about history and literature from contact with the best in our circle, I learned to play chess from another, several repaired their weaknesses in math from being tutored by the best in our circle (me).

However, then I encountered Jensen's analysis of the twin or kinship studies. They showed that cognitive skill differences between individuals were determined largely by their genetic differences and that differences in systematic environment like SES (between family differences) counted for very little: so little that the environmental gap had to be implausibly huge to matter (it would be huge for those clear off the normal scale of environmental quality, but there were very few of those). Inherited brains counted above all, which left only a shadow of environmental potency intact. Whatever environmental factors worked through genes counted: under-reproduction by the bright or hybrid vigor as a factor that matched fewer recessives during sexual reproduction. Whenever environment had a direct impact on the health of the brain counted: adequate nutrition particularly in the womb and immediately after birth (breast feeding) or brain trauma either during delivery or later.

This perspective liquidated all my explanations of the positive manifold except PPC. All of the others assumed that an autonomous or sociological dimension, even a high-culture dimension, was important – and clearly it was not. Jensen had put virtually all of the causal arrows pointing in one direction: from the healthy brain toward conscious problem-solving, with no arrows (at least no cultural arrows) pointing from conscious problem-solving toward the development of an enhanced brain. The combination of the positive manifold *plus* the twins embedded a concept he called "the g beyond factor analysis"; the g that arose from factor analysis was merely our best quantitative measure of it. The combination gave him a theory of intelligence worthy of investigation.

g as the irreplaceable fuel

According to Jensen, the source of the positive manifold was something like this. Deep in the brain's structure, deep enough so that normal environmental differences did not much affect it, there was a petrol station (or a chain of petrol stations) that pumped a certain quantity of neural energy best called true g (rather than measured g) to the various engines of the conscious mind: the engines that did Wechsler's various subtests (searching for definitions, accessing general information, doing mental arithmetic). The only way the engines could perform better was a better grade of *g-fuel*. You could make them run on alternative fuel but then they were merely spinning their wheels. Practice effects did nothing to upgrade *g*-fuel but just made the engine purr: it just produced higher scores that had no real-world significance. So the fully elaboration of Jensen's theory-embedded concept of intelligence was this: *g as the irreplaceable fuel.*

This theory-embedded concept offered a criterion of what was real evidence and what was hollow evidence. The magnitude of score gains over time on the ten Wechsler subtests did not tally with the same subtests ranked by their *g*-loadings; therefore, the gains were not *g* gains; therefore, insofar as they responded to social change, they must be "hollow" with little real-world significance. A fragment of them might be real: gains in response to better nutrition; curing childhood diseases; better pre-natal, peri-natal, and post-natal conditions; and hybrid vigor.

Jensen's theory-embedded concept of *g* accounts for why when we discussed race we were like ships passing in the night. He could not understand why I did not pay more attention to biological factors (those just listed – the ones that entirely dominate his discussion of race in *The g factor*). I had considered them, but I also concluded that he was correct in that they had little potential to explain the IQ gap between black and white. I emphasized that American blacks had a distinctive subculture: one absent in Germany where black and white showed no *g* pattern in their subtest differences (setting aside whether the Full-Scale IQ gap disappeared or not).

Like all honorable men he was very polite. But he delicately hinted that, thanks to the twins, cultural differences of this sort could not really be expected to count for much. Twins, twins everywhere, and (I felt) no more room to think. He did not live long enough for us to discuss the evolutionary scenario he endorsed to explain why genes for the various races were supposed to differ (in terms of what quality of *g* they conferred): the hypothesis that whether or not you endured extreme cold during the ice ages was crucial. I discovered that the inhabitants of the southern half of modern China had ancestors who did not live north of the

Himalayas during the ice ages, and therefore should have been less intelligent than the northern Chinese, whose ancestors did endure extreme cold. They are not.

Jensen and his concepts

Jensen never appreciated the role of meta-theory and heuristics. In *The g factor* (1998), he castigates those who debate the concept of intelligence (with some merit) and yet arrive at no definition that has any specificity or mathematical exactitude. He does not see that a heuristic should not be specific but broad enough to allow competing theories to fit under its umbrella and compete in terms of the evidence.

He says that he is forgoing the term "intelligence" entirely (he does not stick to that of course: to even describe his theory he has to use substitutes like "who learns better or faster"). He will only discuss *g*, which has the required exactitude to serve as a scientific concept. Here he is quite right: every theory has a theory-embedded concept of "intelligence" as a cornerstone and it should be precise and preferably measurable. As we have seen, his concept was *g* as the irreplaceable fuel. How viable, in the light of evidence, is his theory today?

There is nothing logically impossible about Jensen's concept. Some of the body's excretions, like urine, are little affected by culture so long as people are operating under normal conditions of diet. However, many made the mistake that some Newtonians did in the nineteenth century. They would not look at any theory of the heavens that did not embody Newton's concept of celestial mechanics, which turned Newton's theory-embedded concept of gravity into a heuristic: an unalterable concept that gave advice about what sort of theory could qualify as plausible.

Some *g*-centric thinkers elevated *g* to a heuristic. Which is to say that they committed a type-three error. They took the theory-embedded concept of *g* and made it into a piece of *advice*, into a criterion all evidence had to pass to be deemed valid rather than bogus. Thus, any suggestion that culturally induced IQ gains over time were real was suspect – simply because it could not be shown that those IQ gains were g gains. What better way of dismissing falsifying evidence than to label it as no evidence at all. I once gave a seminar in Barcelona to which the response was, "but you have done nothing to show that the gains were on g."

Others sensed a mistake here and began to test whether or not g was really an irreplaceable fuel. The evidence they accumulated was of two sorts: did culturally induced gains, which were not g gains, fuel cognition in a way that made an important real-world difference (is there really an alterative to g-fuel); did certain groups, groups separated by subtest scores that did not tally with g, differ in a way that no one could refuse to describe as an intelligence difference?

Falsifying g-centric theory

I will again elaborate the criterion Jensen (1998) offered to determine whether score differences tallied with g. Take IQ gains over time from one generation to the next: you rank the ten Wechsler subtests in order of the magnitude of the gains on each subtest, and then you rank the same subtests in order of the size of their g-loadings. The g-loading told you the extent to which a particular subtest measured g, in the sense of what subtest was most predictive of the positive manifold: the tendency of a good subtest performance to be sustained over all ten subtests. Unless you found a robust positive correlation between the two hierarchies (biggest gain = highest g-loading, and so forth), the score gains did not constitute a g difference. IQ gains over time generally flunked this criterion and were therefore "hollow." Whatever fueled them was not g-fuel.

Coyle and Pillow (2008) show that the cognitive skills measured by the SAT predict university grades even after g has been removed. Woodley (2012a) shows that education in particular cultivates specialized patterns of cognitive abilities and that these improve independently of whether they correlate with g. Ritchie et al. (2014) are quite explicit: the association of education with improved cognitive performance is not mediated by g; education directly affects specific IQ subtests. Woodley (2012b) shows that the historical trend of IQ gains (which of course are not correlated with g) both parallels and predicts the growth in GDP per capita experienced by Western nations over the last ten decades or so (correlation = 0.930). Meisenberg (2014) argues that over time we are accumulating "cognitive human capital" that is interdependent with economic growth.

There is an inference here I want to defend: that schooling promotes a variety of cognitive skills (g aside) and these promote economic progress. Note that the causal arrows could go in the opposite direction: x causes us to get richer and we spend more on schools and get "smarter." My inference is more probable when we look at "lagged correlations" or what happens when the dimension of time is included. Ireland enhanced education, its tests scores rose, and its per capita gross domestic product rose above that of England – *in that order*. Finland enhanced education of its poorest students and duplicated Ireland's trend (Nisbett, 2015).

Fox and Mitchum (2013) show that IQ gains on Raven's reflect the kind of problems we can solve, despite the fact that they are not correlated with g and are not factor invariant. Fox and Mitchum (2014) extend their analysis to Letter Series and Word Series and show that the fact that the present generation has developed new habits of mind is the very reason gains are not factor invariant. Woodley et al. (2013) conclude that autonomous mental skills allow people to cognitively adapt to modernity and thus score higher on personality indexes. Flynn (2012a) shows that the fact that American adults with some tertiary education went from 12 percent to 52 percent between 1953 and 2007 registered as gains on the WAIS Vocabulary subtest. These were the equivalent of

17 IQ points (over 1 SD). Irrespective of whether the overall pattern of American subtest gains correlated with g, this had real-world consequences: they could carry on different conversations and read a wider range of books. Flynn (2013) suggests how cognitive progress independent of g has enhanced moral maturity (but not political maturity).

Flynn et al. (2014) put the final nail in the coffin. They compared the Wechsler subtests scores of typical subjects with those who suffered from iodine deficiency, pre-natal cocaine exposure, fetal alcohol syndrome, and traumatic brain injury. The typical subjects were higher on every subtest. However, the magnitude of their advantages by subtest had zero correlation with the size of the subtest g-loadings. It is difficult to deny that the typical subjects had a significant cognitive advantage over the four comparison groups. This is not to say that their advantage was analogous to that of one generation over another. The latter was influenced by the new habits of mind that evolved over the twentieth century.

As usual, committing a type-three error (elevating the theory-embedded concept of g to a heuristic) led to a type-one error: refusing to revise a scientific theory despite that fact that there is a lot of evidence against it. There always was something odd about that theory. Two basketball teams are evenly matched. The coach of one decides to drill his players on the fundamentals, layups, and foul shots, simple tasks that are less "basketball-g" loaded. Therefore, the performance gains they make do not correlate with a hierarchy of basketball-skill g-loadings (no gains on complex tasks like fade-away jump shots). Yet there are real-world consequences: his team beats their rivals by ten points.

Bill Dickens and solving the stellar parallax

But must we leave unsolved the absence of a stellar parallax? Do we have to say: "how curious that one striking piece of evidence (the twin studies) seems to show that it is wrong to move on to a new theory when so much suggests that we should." It is time to discuss the assumption hidden behind Jensen's interpretation.

The assumption that lay behind the absence of a stellar parallax was always explicit, that the stars could not be that far away, and it took only a feat of the imagination to call it into question. The Jensen model had a hidden assumption that was so deeply buried that it took Bill Dickens to make it explicit. We will elaborate the Dickens/Flynn model in the next chapter, but briefly: first, Dickens posited that genes and environment simply become more highly correlated as we age, which implied that their influence was additive and not a matter of one draining potency away from the other – the potency of the environment was *masked* by the combination, which thanks to the twin studies was ascribed to genes alone. Second, Dickens posited that current environment eventually obliterated the effects of past environments so that at maturity we should not expect to see traces of the time when the two were uncorrelated – current environment had only a feeble memory of past environments except under unusual circumstances (like brain trauma).

To set the record straight about the Dickens/Flynn model, Dickens had these two insights and modeled them. My contribution was to supply labels and catch an error that led him to go on to invent the social multiplier as a key to the potency of IQ gains over time. I also insisted on a sports analogy to convey how the model affected the real world beyond IQ – he supplied basketball (I had suggested track and field which would not have been nearly as effective).

The result of his insights was, of course, to make explicit Jensen's hidden assumption. Jensen assumed that as we age genes and environment play a zero-sum game. A zero-sum game is one whose rules dictate that anything one player gains the other player loses. Thus, when the genetic portion of IQ variance rose, that *must* mean that whatever genes gained in potency, environment had lost. Dickens and Flynn deny this: when genes and environment become more and more correlated, the potency of one is added to the potency of the other and does not simply disappear.

Try an image from Plato's Chariot. Two horses draw a chariot, an obedient horse and a wayward horse, and as long as the two are at odds they both affect its path. Slowly the obedient horse gains dominance and the two horses pull together. This is what happens as an individual ages. Gene-influenced performance slowly "attracts" an environment of equivalent quality, and environment loses it independence: it cannot do much to give you cognitive abilities uncorrelated with your genetic potential after the age of 20. Except that the chariot is really a troika: there is also a third or "chance" horse that the mature individual can use to override genes to a modest degree (gain points on Wechsler Vocabulary).

But if you think the wayward horse has lost its potency, see what happens when the obedient horse is unshackled. How potent environment can be when freed from capture by genes is shown by massive IQ gains from one generation to another. This is represented in the model by triggers that set the social multiplier to work. Progressive modernity leads to more and better formal schooling, which becomes a feedback mechanism (every person who gets more credentials raises the bar so everyone wants more credentials, and so forth). And more formal schooling (among other things) gives us the new habits of mind to master Raven's.

In sum, Dickens freed me from the spell that cognitive ability was impervious to social environment (except in extreme cases – a child whose environment fell off the bottom of the normal curve) and must be rooted deeply in the brain, subject mainly to factors that had a direct effect on the brain: genes, pre-natal environment, birth trauma, nutrition, hybrid vigor, all the factors that Jensen gives precedence in *The g factor*. I was free to return to my pristine reaction to why *g* exists – that is, why people who tend to do well on one cognitive skill tend to do well on others.

There is no single factor that lurks behind the various high cognitive performances that high-IQ people exhibit when doing the various Wechsler subtests. There is a mix of causes. Now that I know a bit more perhaps I can describe them more fully. There are genetically influenced physiological factors (PPCs) that affect all kinds of complex problem-solving (optimum capacity to generate neurons, connections between neurons activated where they are used, optimum dopamine sprayers that thicken used connections, optimum blood supply to all areas of the brain). There is the fact that exercising one problem-solving skill has a functional connection to exercising another (a bigger vocabulary almost always means reading more widely and accessing a larger store of general information). And there is the company we keep. All sorts of institutions from family (until maturity), friends, leisure companions, and work bring (say) larger vocabulary types into contact with one another and challenge their memberships to think better in many areas, sometimes even mathematics.

I am now in a position, I believe, not only to question the g-centric theory of intelligence but also to make explicit the outlines of a meta-theory that is emerging in its place, particularly since the new theory is emerging partially thanks to a phenomenon that the old theory could not accommodate: massive IQ gains over time.

Elaborating the current meta-theory

Like physics during the Einstein versus quantum-theory era, the current meta-theory divides the study of intelligence into several areas and offers quasi-heuristics (second-level heuristics) in each. Unlike astronomy, where it is easy to estimate the mass and speed of a planet and afford the law of gravity an exact prediction in all cases, it is very difficult to quantify the cognitive quality of an environment or the social changes that trigger IQ gains over time or the structure of the brain. This does not keep us from trying: books in the home, more years of schooling and cognitively demanding jobs, fewer children, estimating the size of brain areas before and after exercise, and so forth. First, I will state three "area-specific" heuristics. I think they offer good advice and thus have allowed scientific theories to emerge that are fruitful: theories that use comparative data and models to generate quantitative predictions.

1. Individual differences within a cohort

Heuristic: *fashion whatever instruments best compare individuals for the cognitive skills their culture emphasizes.*

It is important that when scores on these instruments are taken into account, they add to predictability of university academic performance – that is, add predictability to what we get from a person's academic record alone. Within the context of modernity or near modernity, my favorites are the Stanford-Binet, the Wechsler tests, Sternberg's tests, and the Woodcock-Johnson. Pre-modern societies need their own tailored instruments whose "subtests" sample the cognitive skills valued.

Almost immediately there arises the problem of what genetic and environmental factors promote good performance, with a premium on distinguishing environmental factors correlated with genes from those that are uncorrelated, and subdividing the latter into "pure" chance and a sphere of personal autonomy (both of which can put you above or below your genetic promise to a significant but limited degree). As for non-cognitive factors that promote competence, these are legion and debate about how many are important will be summarized in the next chapter on competing scientific theories (all the way from *g* theories to Gardner's multiple intelligences).

This book generates predictions such as that Americans in a humdrum job and social circle (but blessed with the competence to become mature students) could gain as much as

11 IQ points. Comparative data show that Israeli females in highly orthodox homes would gain about 8 points on Raven's if allowed access to modernity, and that low-SES children out of school over the summer stall in terms of enhanced cognitive maturity. The Dickens/Flynn model predicts that intervention programs must alter character and the quality of after-intervention peer groups if IQ gains are to be sustained. After the intervention is over, later environment will swamp the intervention environment unless the aging person, or their peers, generates sustained environmental quality. Any reader can list dozens of fruitful predictions.

2. IQ trends over time

Heuristic: *these are dictated by altered social priorities that affect the cognitive problems habitually confronted and deemed worth solving.*

During the twentieth century, these priorities and habits of mind have changed radically as societies begin to industrialize and enter the world of modernity.

Comparative data suggest that when a nation goes from pre-modern to full modernity, it will gain at least 36 IQ points, usually more than that on Raven's, which emerges as the best measure of progress toward modernity. As for real-world significance, that IQ gain is part of an interactive process that alters a pre-industrial society from what it is today to something like what we are today. We are trying to quantify how modernity's new habits of mind and new levels of problem-solving (well removed from the concrete) affect everything from IQ test performance, to academic performance, to economic progress, to the emergence of democracy, to human happiness. We should be correlating just how the pattern of gains on Wechsler subtests mirror just what aspects of modernity a nation has achieved thus far (as in the Sudan).

Generations rarely differ by any appreciable degree of genetic quality, although there can be minor differences due to selective migration, selective reproduction, hybrid vigor, and cataclysmic events like the extermination of an elite.

However, there are group differences other than those between successive generations: for example, differences between ethnic groups within a society that vary by subculture. The latter are unlike generational differences in that the possibility that genetic as well as environmental differences possess relevance is much more real. Quantification of the effects of subculture on IQ is difficult. An exception (based on a small sample circa 1980) was Elsie Moore's estimate that black American subculture had a cognitive quality that cost children 13.5 IQ points by the age of 8.5 years. That deficit was present even when two groups of adoptive parents (black couples versus white couples) were matched for maternal years of education and roughly for elite SES; and all of the children adopted were black (genes thereby controlled).

Social classes are a case in which both environment and genes are relevant (unless you are mad enough to deny that social mobility is affected by intellect or that genes have an effect on intellect). Even here we can sometimes quantify evidence to test a thesis. It has been suggested that in advanced societies the hierarchy of genetic quality and the class hierarchy have become more highly correlated. If so, the IQ gap between children of the top third in terms of occupational status and the bottom third should be increasing. Yet American and other data show that it has been stable at about 10 IQ points.

3. Brain physiology (the PPC, or physiological prerequisites of cognition)

Heuristic: *the brain is like a muscle in its plasticity but organized like a system that is both decentralized and federal.*

By that I mean that when cognitively complex tasks are done, although much of the brain is involved, the circuitry varies and some areas are more prominent then others.

In all sports, from weight lifting to swimming, most of the body is involved but it is coordinated in different ways with more stress on some muscles than others. The body possesses common factors like its peculiar capacity for developing muscles (analogous to the potential for the multiplication of neurons area by area), the quality and maintenance of the links that coordinate them (analogous to the quality of the dopamine sprayers that "thicken" connections between neurons), and the cardiovascular system (analogous to the blood supply that feeds that whole brain). But various muscle groups are more developed by different exercise as the muscles of the weight lifter and swimmer show (and areas of the brain differ in terms of exercise: for example, the special relation between mapping and the size of the hippocampus).

Magnetic resonance imaging and new technology promise much more than the above crude "map" of the structure of the brain. Physiologists have begun to quantify "the integrative framework" of the brain that underlies complex, goal-directed (executive) behavior. Many are quantifying the effect of mental exercise on various areas of the brain, such as map reading on the hippocampus, or playing video games on the cortex.

Hopes for unification

Just as physicists today are trying to integrate sub-atomic and super-atomic physics, psychologists are trying to unite the three areas of the study of intelligence. Many believe that the best hope of unification at present would be a theory-embedded concept of *executive function* that would encompass the three areas of individual differences, trends over time, and brain physiology. By that I mean they are turning toward working memory as a process that enables one to hold goal-relevant information in mind, even in the face of competition from other kinds of cognizing and despite distraction (from, say, emotional interference). However, these advocates have a long road to travel.

First, they will have to develop mental tests focused on measures of working memory (with perhaps Vocabulary, Information, and Arithmetic as add-ons) and show that these do better than current IQ tests at predicting individual differences in cognitive performance, SAT scores, university grades, qualifying to do cognitively demanding jobs, who escapes mental retardation, and so forth. The Wechsler battery has increased its working memory content (letter-number sequencing). Let us see if that constitutes progress.

Second, focusing on largely pre-modern nations they will have to measure their progress using the new tests – all we have are gains on conventional tests. They will have to give them both kinds and show that theirs correlates better with growth in GDP per capita, academic skill gains, and the trappings of modernity (democracy).

I am skeptical in that I believe the shift from pre-modern to modernity is more complicated than this. I suspect that we are no better than our ancestors in terms of holding goal-relevant information in mind. But if we are, it may just be that modern society imposes a wider and more complex array of cognitive tasks and we have had to discipline our minds in the direction of non-distractibility. Also, can working memory capture a crucial psychological factor? We ask a pre-modern man what fish and crows have in common, and rather than saying, "they are both animals," he says, "nothing: you can eat one and not the other." Why has he got this Similarities-type item wrong? I doubt it is because of poor working memory. It is because he has no habit of using non-pragmatic abstractions to classify concrete particulars. Raven's or Piaget seem well suited to capturing the new habits of mind of modernity. A better test will have to be very good indeed.

Among adults Vocabulary has shown great increases since 1950. Could working memory explain these, or would we need to take the rise of years of formal education into account to explain the peculiar gains on Vocabulary? The size of gains on various Wechsler subtests varies greatly and I cannot believe that any one conceptual skill can explain the variation.

Third, imaging research along with lesion and trauma research must identify the brain processes that maximize excellence in working memory – and show that these processes described in their own terms are better predictors of cognitive performance than alternative maps of the brain. I think the first will be easy, the second difficult.

There is an underlying reason why total unity may never be possible. The psychological profiles of people doing tests are simply not comparable when we are measuring individual differences, group differences, and generational differences; worse, a psychological or neural comparison cannot substitute for a sociological dimension. Take four comparisons. We can contrast the skill profiles and brain images of two women and not know that one is still practicing law and the other has retreated to the home for child minding. We can compare male and female university students and be impressed that the former have a 2- or 3-point IQ advantage on the latter; until we realize that this is only because the former are a more elite sample from the general population than the latter (females qualify for university with lower IQs than males). We can do the same for whites and blacks and find that a variety of contrasting cognitive competencies exist because, while both groups are exposed to modernity, the latter come from a cognitively restricted subculture. We can do a skill and neural profile on two people who score high and low on Raven's and not know whether the low-scorer has come from a culture that is pre-modern or is someone who lacks the mental ability to take advantage of modernity.

In other words, the various psychological gaps are so different it is hard to see them being reduced to one. But more daunting, everyone's cognitive behavior is influenced by both psychological and sociological factors and the information conveyed by the latter is both essential and can never be conveyed by the former. I believe we will have to struggle on with three areas rather than one.

The next century

I hope we have learned three great lessons. Do not try to provide a narrow definition of a heuristic (don't waste time trying to define "intelligence"). A heuristic is not a precise concept but a broad piece of advice about theory building. Do not elevate a theory-embedded concept of intelligence to the status of a heuristic (what happened to Newton and to g). And, hardest of all, when measurement tells you something is impossible – the earth cannot move because the stars cannot be that far away, or that intelligence cannot move in response to culture because twins show how omnipotent genes are – try to make explicit all hidden assumptions. Nothing showed that the size of the universe was known; twins did not show that genes and social environment play a zero-sum game.

"Intelligence" has many uses: as a heuristic, to refer to a variety of theory-embedded concepts, and as a sign that one is capable of solving the problems posed by everyday life (does not suffer from mental retardation). But the notion that it is one thing dies hard. Sometimes I tell people the four changes that occurred during IQ gains over time: (1) Thanks to new exercise (habits of mind) we can solve a wider range of problems than our ancestors; (2) That new exercise means we die with a brain differently developed than theirs; (3) But they were not born with worse brains; and (4) They were capable of solving the problems with which their time presented them. And yet, some keep asking: "but are we more intelligent than they were?"

Over the last 100 years, the study of intelligence has developed a meta-theory that promotes good science. It has three levels and three areas: at the top is a good general heuristic; at the next level it has three good quasi-heuristics; these divide it into three areas which, thus far, allow for a variety of fruitful models and predictions, and have generated a huge body of quantified data. In other words, we do not need to replace these heuristics and are not committing a type-two error in hanging on to them. I know that my three concepts are humble: a predictive measure, shifts in habits of mind, neural federalism. I concede that they do not have the precision to which Jensen and the champions of working memory aspire. They are broad, which is precisely what they should be: broad enough to play their role of giving appropriate advice in each of the three areas of intelligence research.

In the next chapter, I will try to show that these heuristics have guided the formation of a wonderful variety of scientific theories. The quality of these theories reflects credit on the heuristics we have today.

Science and social science

The fact that we have two levels of heuristics (one overall guide plus three area guides) is typical of a social science. For example, international politics does best with three concepts that guide theory building: calculation of national interest, affinities with other nations, and a nation's historical narrative. Unifying concepts have proved a hindrance. Quincy Wright tried to reduce the discipline to one theory alone. All national behavior was to be organized in terms of twelve sets of coordinates, which he described as analogous to maggots eating their way through multi-dimensional semi-opaque cheese.

The concept of g as a unifying concept led us astray; let us hope that working memory does not play the same role. I anticipate another century of scientific progress. Someday, an unexpected source of data may emerge, and we will need new theories or even a new heuristic. Until then, enjoy the sun while it shines.

If we keep doing good research we should not quibble too much about whether we are really doing hard science. Some years ago at Otago, a lecturer suggested that departmental meetings be held about how to turn psychology into a science. After each fraught meeting, the chair was seen ashen, ordering an unaccustomed drink before lunch. In the end, they adopted a resolution that they should give more tutorials for students at the second-year level.

Answers

(1) The historical record shows that it is a mistake to use a precise theory-embedded concept to guide or unify intelligence research.

(2) We should make do with a heuristic and sub-heuristics broad enough to allow theories to freely compete at a "lower" level. Their breadth allows them to generate competing hypotheses about what traits predict success within a culture, what social changes alter our minds over time, what neural pathways are energized when we solve certain problems, and so forth.

10   Scientific theories of intelligence

Questions

(1) Do current theories of intelligence "fit within" with my own meta-theory?

(2) How compatible are these theories with one another?

We abandon meta-theory to discuss theories in the narrow sense of scientific theories. These undertake the strictly scientific task of explaining phenomena under three headings: clarifying the nature of individual differences, or group differences (including differences between the generations), or brain physiology. I do not believe that any of them challenge either my overall heuristic for intelligence or the quasi-heuristics that give advice to scientific researchers in the three areas. They do not all fit neatly within one of the three areas. Indeed, it would be a poor theory that did not have implications for one of its neighbors. But usually, they originated in one area and I will classify them by their area of origin. The exceptions are theories that were actually designed to bridge areas – that is, were attempts to reconcile findings within two areas that seemed incompatible with one another.

Individual differences

I. g-centered theories

Jensen

I have discussed Jensen's theory at length as a counterpoint to developing my own and will be brief. Its salient feature is its focus on *g*, the general factor that emerges from the fact that cognitive abilities are inter-correlated (people who do better on one tend to better on all). Jensen was quite aware that factor analysis produced other cognitive abilities that were also significant: verbal factors, memory factors, and so forth. For example, he argues that black and white Americans are relatively equal for rote memory while unequal for more complex thinking.

But even so, aside from race, he did little to analyze the operational cognitive abilities tested by the various Wechsler subtests – Vocabulary, Arithmetic, Comprehension, and Information, the abilities that have so much to do with the significance of individual differences. For example, less vocabulary means less reading, less adequate performance on the SAT, and so forth. I do not mean he would deny the latter. But once he had determined the flow of *g* (the irreplaceable fuel) through the subtests, he seemed to lose interest in them in their own right. This may well be a psychological consequence of his theory rather than a logical consequence. I see no reason why a *g*-man should not have done the research contained herein such as measuring the role of the family in creating an injustice through its influence on Vocabulary.

My major objections arise on the level of group differences. Jensen virtually defined the significance of IQ gains from one generation to another out of existence: his demand that they must be either *g*-differences or hollow. But concerning group differences between black and white Americans, the fact that black disadvantage rises with the cognitive complexity (or *g*-loading) of the Wechsler subtest posed an interesting problem of explanation: was this a matter that black genes were more disadvantageous for complex tasks, or was black subculture such that it discouraged the development of cognitively complex skills? Jensen favored the first alternative; I have argued strongly for the second (Flynn, 2008, Chapters 2–4).

Despite my reservations, Jensen made contributions that are still relevant: g serves as a measure of the cognitive complexity of various cognitive tasks (their g-loadings). We might have suspected that digit span backward (repeating numbers in reverse order) was more complex than digit span forward (repeating numbers simply in the order in which they were read out), but what of the relative cognitive complexity of vocabulary and simple mental arithmetic? The hierarchy of g-loadings makes ranking tasks for cognitive complexity less of a guessing game.

On the level of brain physiology, there is evidence that inbreeding is more deleterious the greater the g-loading or complexity of a cognitive task. This implies that bad luck in sexual reproduction (happening to get two damaging recessives aligned) affects the brain more the greater the cognitive complexity of the task. Which in turn poses this hypothesis: those areas/networks of the brain that are the seat of complex thinking are more vulnerable than those that are the seat of less complex thinking (say, the hippocampus as the substratum of map reading).

Cattell-Horn-Carroll (CHC theory)

As a thinker, Cattell should be read selectively. He founded a religion based on Social Darwinism, and suggested that American blacks be confined to reservations to be treated kindly if they agreed to be bred to extinction – he called this "genthanasia" (Flynn, 2000). In psychology, Cattell (1941) distinguished between: fluid g or the ability to do on-the-spot problem-solving (not dependent on prior knowledge), the sort of items on Raven's Progressive Matrices; and crystallized g or the sort of knowledge an intelligent person tends to accumulate, such as vocabulary and information. He saw fluid *g* as an *investment* that paid dividends in the form of crystallized *g*, all of the skills that learning affords us in dozens of specific areas: not just vocabulary and information but every mental skill with a cognitive content that we gain from thinking our way through school and life. This is similar to Jensen's view of *g* as an irreplaceable fuel.

Chapter 7 shows that this fluid skill is just as heavily influenced by family environment as the most malleable crystallized skill (vocabulary) and therefore, neither skill deserves to be called an investment and the other a dividend. The presumption is that they develop by mutual causality: mental acuity promotes more vocabulary acquisition, and more vocabulary acquisition (through reading cognitively complex books and talking to cognitively acute peers) promotes more mental acuity.

However, Cattell's distinction is important when abilities decline, as distinct from when they are acquired. Performance on Matrices tends to decline starting about age 25 and vocabulary begins to decline about age 55, so something has now made the two functionally independent. Clearly the analytic areas/networks of the aging brain begin to deteriorate long before the verbal areas/networks do, so the crystallized abilities become self-sustaining with continued use even though the fluid abilities do not. I have tried to analyze these trends in terms of the four kinds of abilities factor analysis derives from Wechsler IQ tests. I have argued that in old age, those of high analytic ability decline faster than those below them (pay a bright tax), that those with high speed of information processing are the same, that those of high verbal ability decline slower than those below them (get a bright bonus), and those of high working memory are neutral in this regard (Flynn, 2012a). These hypotheses are based on cross-sectional data and must be tested by longitudinal studies (actually tracing how the abilities of individuals change as they age).

Thanks to factor analysis by John Horn (1965) and John Carroll (1993), with supplementary development by McGrew (2005),

Schneider and McGrew (2012), and Flanagan et al. (2013), Cattell's insight evolved into a three-level theory:

(1) At the top is simply *g* without subdivision.
(2) Immediately below are ten broad areas, fluid intelligence (*Gf*), crystallized intelligence (*Gc*), quantitative reasoning (*Gq*), reading and writing ability (*Grw*), short-term memory (*Gsm*), long-term memory akin to working memory (*Glm*), visual processing or analysis of visual patterns (*Gv*), auditory processing (*Ga*), processing speed or speed of information assimilation under time pressure (*Gs*), and reaction time or how quickly you react, measured in milliseconds, to a visual or auditory stimulus. Some have proposed adding to this list.
(3) At the bottom are over seventy narrow abilities that are highly specific areas of knowledge, having to do with science, culture, geography, mathematics, number facility, reading, spelling, grammar, writing, vocabulary, fluency, general information, listening ability, induction, memory, attention control, naming, visualization, spatial scanning, coding, perceptual speed, and many more.

From my point of view, this is all to the good in that factor analysis is now emphasizing a variety of cognitive abilities. However, we must view them through the spectacles of the sociological imagination and abandon investment theory. By this I mean that trends over a person's lifetime and trends over generations in the narrow abilities can occur without being in tandem with their rankings on a *g* hierarchy (see Box 4). And their significance for the individual's prospects and societal cognitive progress is appreciated in their own terms: whether your vocabulary gets you into a good university is important, *g* aside. If you think these "lowest level" abilities can increase only to the degree that more *g* is around to be invested, that would be an inhibition.

Box 4   A valid investment hypothesis

I should add that I accept another kind of investment hypothesis, which is independent of any particular scientific theory, namely: if you invest heavily in the development of some cognitive skills (say, verbal), you may invest less in other cognitive skills (say, mathematical). It would be odd if this were not true: people have only a finite amount of time and energy.

In terms of actually measuring individual differences, the Woodcock-Johnson tests have capitalized on CHC theory. The fourth edition attempts to measure g and ten intermediate abilities, of course, but also thirty-five narrow abilities (Flanagan, 2014). I have been unable to locate studies of whether the Woodcock-Johnson better predicts university performance than the Wechsler and Stanford-Binet tests. Unlike the latter they are not individually administered.

Ackerman (1965) developed a theory based on Cattell called PPIK theory, standing for process, personality, intelligence, and knowledge. Although it retains the notion of investment of cognitive ability in the pursuit of knowledge, it gives personality a much larger role. Task-oriented people generally reflect about problems, a trait less pronounced among "active" types (inclined toward physical strength and aggression), and "artistic" types (inclined toward self-expression). The knowledge accumulated by these three types may not overlap. Research to verify this typology has an affinity to Bandura's hypotheses (see below).

II. Sternberg

Sternberg has made an impressive attempt to broaden the content of tests like the Wechsler and Stanford-Binet, so that we can measure a wider range of abilities that allow us to adapt to the cognitive demands of our time (Sternberg, 1988). He originally called his theory the Triarchic Theory of Intelligence but has renamed it the Theory of Successful Intelligence. He argues that it best predicts success in life from the point of view of the individual, operating of course in the relevant social context (Sternberg, 1997). It gives the individual practical advice about how to play to one's strengths, not only how to adapt to one's environment but also how to select and shape one's environment.

Sternberg acknowledges that the usual tests are a good measure of g but argues that g has exhausted its scientific potential even on the level of individual differences. He calls g "the academic form of intelligence" and believes that it falls within only one of three important competencies:

(1) His *analytic intelligence* measures something close to fluid g – that is, solving abstract problems on the spot as in Raven's.

(2) *Creative intelligence* tries to go beyond Raven's to test on-the-spot creativity of a less cerebral sort: for example, selecting cartoons with blank captions for the characters and filling in what would be appropriate and clever, or writing impromptu stories on themes like the octopus's sneakers.

(3) *Practical intelligence* is an attempt to measure skills used to apply concepts to real-world contexts: for example, how to deal with writing a recommendation for someone you do not know well, handling a competitive work situation, or how to deal with a difficult room mate. Its measurable core is tacit knowledge. The latter is very close to the capacity of Aristotle's person of practical wisdom to find the golden mean between two extremes. Some people, whether by nature or by habituation, are much better than others at determining what ought to be done on the battlefield, rather than being too cautious or too rash.

Sternberg emphasizes that the traditional tests have had a century to accumulate studies that attest to their external validity (ability to predict individual performance), while his tests have a history of less then three decades. He cites studies that seem to show that his test betters g as a predictor of real-world performance in work situations (Sternberg et al., 2000). His most impressive achievement has to do with prediction of university grade point averages (GPAs). By adding his three measures to the traditional predictive variables of high school grades and SAT scores, he increased the percentage of variance explained from .159 to .248 (Sternberg, 2006). Which is to say that the correlation between the predictive measures and university grades increased from 0.40 to 0.50.

Jensen (1998) was highly critical of Sternberg's measures. It seems obvious to me that they do measure a new range of relevant skills, such as tasks that would help to predict things like how interesting a student's essays will strike university staff.

III. Gardner

Gardner (1983) offered a list of seven intelligences:

(1) Linguistic. Mastery of the meaning of words and the syntax of language, with an ear for sound and an eye for imagery important for those few who become stylists or go on to write literature or poetry. Both they and rhetoricians must be aware of how language affects emotions.

(2) Logical-mathematical. He emphasizes that mathematics involves more than logic, such as the capacity to entertain long chains of logical relations expressed in symbolic form.

(3) Musical. Performing music, which in a small number of cases leads on to musical composition, although composing can begin at an early age.

(4) Spatial. Spatial visualization, seeing the continuity of a shape being rotated in space, and the power to create

a mental image, which when highly developed is useful in mathematics and chess.

(5) Bodily-kinesthetic. This affects all areas in which control of the body or exploitation of its potential are central, such as sport, dance, mime, and acting. Later, he clarified this intelligence by saying that it is the very bodily skill of the athlete or dancer or surgeon that earns the accolade, and stressed the enormous amount of practice and expertise that goes into their performances (Gardner, 1999, pp. 95–6).

(6) Self-oriented personal intelligence. A sense of personhood, self-knowledge of one's own feeling, capacities, and limitations, and control over one's behavior.

(7) Other-directed personal intelligence. Knowing others in a way analogous to a mature knowledge of self, culminating in the kind of empathy that characterizes good teachers and therapists and great leaders. Note that these personal intelligences are not equivalent to mere sociability but, rather, are forms of people-knowledge.

Ten years later Gardner (1993, p. xviii) added an eighth kind of intelligence called naturalistic intelligence. This refers to those expert in discerning the flora and fauna of their environment and those who go beyond nature to recognize automobiles by their sounds, discern artistic styles, and see novel patterns in the laboratory (Gardner, 1999, pp. 48–52).

Gardner is mainly concerned that the eight abilities all be called the same name, no matter whether it is "intelligences" or "talents." This is because he believes that the two terms constitute a value hierarchy. Calling what IQ tests measure (linguistic and logical-mathematical competence) intelligence and calling excellence at dance a talent devalues dancing. It implies that those who lack IQ skills but have outstanding kinesthetic or musical abilities are not smart but dumb (Gardner, 1983, p. xi; 1993, p. xx).

Gardner has been attacked on theoretical and practical grounds. The theoretical issue is what level of cognitive complexity a trait must meet to qualify as a kind of intelligence. Gardner (1983) defends his list by arguing that all of them solve problems that progress from the elementary to the advanced, that they have a physiological substratum, and that they are autonomous in the sense that individuals rank higher on one than on another (and thus the abilities do not inter-correlate to a significant degree). Stenberg counters that, nevertheless, would we really count an adult who is tone deaf and has no sense of rhythm as mentally limited in the same way we would someone who can never learn to talk?

When I look down the list, the abilities fall into two categories. First, there are those that involve a high degree of cognitive complexity such as linguistic, logical-mathematical, and spatial, the abilities that are measured by orthodox IQ tests like the Wechsler subtests. In terms of cognitive complexity, we might rethink musical creativity: the complexity of the "design" of a Mozart symphony may be as great as Einstein's theory of relativity. It is certainly extraordinary that Mozart could hold that design in his mind as a simultaneous concept, while most of us have to hear it unfold over time. Perhaps when we know more about bodily-kinesthetic ability, we will find it has a larger cognitive content than it is traditional to assume: a boxer has a strategy, a basketball point guard instantaneously maps just where everyone is on court, and so forth.

Second, there are personal traits that must accompany cognitive abilities if they are to be operational in important areas: self-knowledge and knowledge of others. As we have seen, Sternberg counts these as important in the area of practical intelligence, and they can be amplified into Bandura's emphasis on motivation, self-control, a sense of self-efficacy, and an awareness of the consequences of the exercise of ability in a social context.

The practical issue is my concern – namely, Gardner's assertion that all of his eight abilities should be called the same name, otherwise calling one an "intelligence" and another a "talent" implies a value hierarchy (puts mathematical competence above dance). This ignores the distinction between whether a hierarchy ought to exist and whether it does exist. Every society has a hierarchy of priorities that evolves over time. For example, in 1900 memory may have played a more important role. Today, there is no doubt that modernity and its job market place a greater premium on the abstract, the hypothetical, a large vocabulary, and mathematical skills. I have never known of a society that gave priority to cutting your own hair at maximum speed. There may be a strategy, practice may help, there is certainly a physiological substratum, you can rank people, and that ranking may not correlate with anything else.

Gardner defends his list by asserting that all of its abilities are socially valued. However, the question is, to what degree? I believe that our society is far too narrow in what human traits are valued. Aristotle says that society is more than a market because you can do business with foreigners, more than a military alliance because you can negotiate mutual defense treaties with foreigners, more than marriage ties because you can marry a foreigner, more than physical proximity because two groups can occupy the same city and be divided by hate, and more than abstaining from injury to others because one can be kind to foreigners. The foundation of civil society is a shared way of life, rich in philosophy, art, sport, amusements, and diversity, whose consummation is a sense of personal loss if anyone else suffers the deprivation of non-participation. It allocates benefits and duties fairly. It develops the full potential of its citizens, unlike dwarfed societies that cultivate only the entrepreneurial (Carthage) or military virtues (Sparta).

In other words, I am attracted to the notion of a society that values a wide range of abilities. If Gardner helps us get better scientific knowledge about all of the abilities he lists, all to the good. If those who have constructed tests based on Gardner give more children useful information about a wider range of talents than traditional tests, all to the good. Excellence at sport or music has saved the morale of children who are not academically gifted. I wish I could make materialistic societies value local theatre and art and dance and music, the amateur athlete, the person who makes a room seem warmer just by entering it, as much as those who excel in making money. But to earn a decent livelihood at most of these skills, you have to be virtually one in 10,000. You will not fool any parent by sending home a report that Johnny may be at the 67th percentile for baseball, but sadly cannot keep up in reading and math. They know full well that bodily-kinesthetic "intelligence" is far less valued than linguistic and logical-mathematical "intelligence," and that you have not improved his prospects by telling him he is bright rather than dumb. Any teacher that confuses parents about what skills may benefit their offspring is culpably remiss.

Gardner has written a perceptive response to my views (Flynn, 2009; Gardner, 2009). He wants full knowledge both on the level of science and communication. He stresses certain dividends of this theory that I endorse. Thanks to his use of the phrase "multiple intelligences," I suspect that schools pay more attention to individual differences, and target their attempts to educate with greater accuracy. Thanks to that phrase, scientists investigate a wider range of mental skills. My message is directed only to those who need to be told that there are no shortcuts toward making American a more humane society. Using an honorific label that obscures America's real hierarchy of social priorities is not the answer.

IV. Bandura

His theory took what most scholars assumed, that cognitive and non-cognitive factors interact, and made the connection explicit (Wood and Bandura, 1989; Bandura, 1993). He argued that cognition evolves in a context that includes both other personal traits – not just motivation but also self-control (executive functions that include the ability to control one's emotions) – and life experiences. For example, family may not provide examples of short-term emotional control in pursuit of long-term goals. Life experiences determine one's sense of self-efficacy, which is conditioned by how well problem-solving seems to work in terms of welcome or unwelcome consequences.

The person's "image" of cognition is modified by prediction of outcomes and selection of what methods will maximize satisfaction. In other words, what kind of cognition is likely to be the most successful? If displays of academic intelligence make you unpopular, it may have low priority. Two students can have the same level of knowledge and skill and yet perform well or badly – this is, may perceive a situation as either attractive (pleasing your teacher and parents) or challenging (the personal threat of alienating your peers). Needless to say, the accumulation of knowledge will be affected, not just academic knowledge but also out-of-school knowledge (when a boy takes up baseball rather than ballet).

A number of scholars have attempted to quantify the roles of non-cognitive versus cognitive factors. Duckworth and Seligman (2005) gave 164 American children an IQ test at the beginning of the eighth grade (age 13). They were also given a dollar bill in an envelope: they could either open it or give it back unopened a week later and get two dollars. The results show that the children's capacity for self-control has twice the weight of their IQs in predicting their grades.

Kelley and Caplan (1993) found that the members of Bell Laboratory research teams all had high IQs. But what distinguished star from average performers was not still higher IQ but effective interpersonal strategies. Heckman and Rubenstein (2001) compared dropouts who qualify for a high school diploma by way of a general educational development exam (GEDs) and high school dropouts who receive no diploma whatsoever. Although the GEDs had higher cognitive skills than the dropouts who earned no GED, they earned no higher wages because they had lower non-cognitive skills. Heckman et al. (2006) have shown that non-cognitive factors, like self-esteem and the degree of control people feel they have over their fate (their sense of self-efficacy), are just as important as cognitive skills for a whole range of outcomes including teenage pregnancy, smoking, marijuana use, and criminal behavior.

By making the relationship between cognitive and non-cognitive factors explicit, Bandura paved the way for other theories that emphasize these interactions. For example, PASS theory stands for planning, attention, and simultaneous plus successive processing. It emphasizes the roles of strategy, alertness, and encoding, transforming, and retaining information (Das, 2002; Das et al., 1994).

Trends over time

I. Flynn

Flynn is convinced that IQ gains over time are symptoms of real cognitive trends between the generations. IQ gains do not directly measure those trends but they supply rough estimates of their relative magnitude. Undoubtedly, they miss certain trends about which we can only speculate. IQ gains must be interpreted sociologically to discover their causes and effects, which is not to say that biological factors have not contributed. Better diet and better health obviously contributed in the earlier phase of industrialization, and still do as explanations for the huge Raven's gains by the better-preserved aged. But for most age groups the later phase of industrialization sees sociology swamping biology.

Causal explanation involves three levels:

(1) Ultimate causes are the industrial revolution and the resulting trend toward modernity.

(2) Intermediate causes are the effects of industrialization on society, more education, emancipation of women, smaller families (with a better adult-to-child ratio), more cognitively demanding jobs, more cognitively demanding leisure, and a new pictorial and symbolic world from television and the internet.

(3) Proximate causes have to do with how people's minds altered, so that in the test room they could do better when taking IQ tests. For example, formal education freed people from the concrete world, which rarely demanded logical analysis of abstractions. They became habituated to take the hypothetical seriously (propositions which refer to imagined states of affairs), to classify things by using abstract categories, and to use logic to analyze abstractions, so as to perceive relevant similarities and differences.

Clearly there is reciprocal causality here in terms of cause and effect: formal schooling encourages forsaking the concrete for the abstract, and those who learn the new habit will be better suited to profit from formal schooling.

This suggests a cause-effect analysis of a variety of IQ tends proceeding from subtest to subtest (Flynn, 2012a). I will address American trends on a few Wechsler subtests:

(1) Vocabulary: more formal schooling, more jobs that require a wide vocabulary, and more peers who do the same. The effects on schoolchildren are slight because they are (say, at age 10) being compared to the last generation of their same age group, and both groups had the same number of years of schooling (four or five). Adult Vocabulary gains are huge thanks to extra education and immersion in a new adult world.

(2) Picture Completion: high gains reflect the new visual culture. An unmeasured consequence of this new world is that it seems to leave less time for reading serious literature

and history. Thus, while Vocabulary gains might seem to foreshadow a more informed public, the net result may be a less informed one.

(3) Arithmetic: tiny gains for both schoolchildren and adults in terms of arithmetic reasoning. We clearly do not know how to teach this skill no matter how long we keep people in formal education. An obvious consequence is the fact that America has to import half of the engineers it needs.

(4) Similarities and Matrices: as we have seen, moving from fixation on the concrete has enormously upgraded our skills for classification (Similarities) and seeing logical sequences in the abstract symbols present in Raven's. Huge Raven's gains are implicit in data from the Raven's test itself, rather than the Wechsler Matrix subtest, which is quite recent. An unmeasured consequence is a break-through in moral reasoning, which puts arbitrary moral principles and racial bias on the defensive.

It seems incredible that a father would kill his daughter for the sake of "family honor" because she had been raped. We would ask, "What if you had been knocked unconscious and sodomized?" But if he sees moral maxims as concrete things, impervious to change, rather than as general principles subject to logic, and sees no point in "speculating" about hypotheticals, he will dismiss your question as totally irrelevant. My father would not have endorsed anything so primitive. But when my brother and I used to challenge his racial bias by saying, "What if you woke up tomorrow with a black skin?" it got us nowhere. He would say: "That is the dumbest thing you have ever said; who has that ever happened to?" He would not take the hypothetical seriously and that is the basis of mature moral argument. Few today would feel they could refuse to show that their ideals were logically consistent (Flynn, 2013).

My approach to IQ gains over time generates predictions. As women are permitted to take full advantage of modernity, they will make gains on men for Raven's IQ (this has already occurred in five advanced nations). IQ gains will end in the most advanced nations (this had already occurred in Scandinavia, Holland, and perhaps France). The intermediate causes of gains are losing their potency. Formal schooling has probably done as much as it can to inculcate the new habits of mind (classifying and using logic on abstractions). Despite featherbedding, we may have created about as many new cognitively demanding jobs as we can. If we are to reproduce ourselves, family size cannot drop much more and, indeed, the solo-parent home is a trend toward a worse ratio of adults to children. If advanced nations make no further IQ gains, and if developing nations still in the first phase of modernity score large gains, the latter will close the IQ gap with the former. This is already occurring (Flynn, 2012a, 2013).

Finally, if the American intellectual elite is no better informed today than in the past, they cannot be expected to restrain the US government when hubris impels it to military interventions in the Middle East (Flynn, 2012b), or unmask its reluctance to come to terms with climate change (Flynn, 2015).

II. Oesterdiekhoff

He is the most original thinker among the continental Piagetians. My account of IQ gains since 1900, insofar as new habits of mind are operative, is an extension of his theory about previous centuries. Oesterdiekhoff (2012) published a seminal article in *Intelligence* that analyses cognitive trends all the way from the most primitive pre-industrial societies to the present. He believes his account explains the evolution of magic into religion into science, inherited mores into humanism, and the rise of democracy.

He takes Piaget's four levels of cognition development and applies them to anthropology:

(1) sensory motor stage: children develop practical and visual skills analogous to animals;

(2) pre-operational stage: by age 2, they can generally develop language reasoning;

(3) concrete operational stage: by age 7, they can use logic to coordinate the concrete world – that is, objects given to the senses;

(4) formal operations stage: they become capable of abstract and hypothetical thinking.

All pre-modern societies are stuck on stages 2 or 3, or have a mixed population with some on one and the remainder on the other. This accounts for their low average IQ. In 1900, no pre-modern or early modern society had a mean above 75 scored against current norms (ancient Greece attained a level in mathematics and science that made it an exception). The low IQs of pre-modern societies dictate cultural traits that resemble those of young children raised in our own society.

The formal operations stage develops only in modern societies, usually sometime between the ages of 15 and 20 (which is close to the peak year of Raven's performance). When this is fully developed, people use logic to analyze concepts – that is, they engage in reflective, abstract, experimental, combinatorial, and hypothetical thinking. This represents Piaget's formal level B or the highest level of cognition, but many do not fully attain it. Even in the most advanced societies, 50 to 70 per cent of adults remain on Piaget's formal level A and are less capable of abstract and hypothetical forms of thinking. If this is so, it tallies with my views on why IQ gains eventually cease on tests like Raven's. Formal schooling (at present) can only do so much to upgrade this kind of cognition, and begins to offer diminishing returns over time or even to lose ground to other social developments that discourage rationality.

When Oesterdiekhoff advocates that whatever cognitive level people attain engenders a wide range of other traits, he means this literally: the traits our own children manifest the child-like worldviews of pre-modern societies. All young children believe that magic is potent, charms and thoughts can kill, dreams are real, animals are like humans, and in the personification of the forces of nature (that mountains, rivers, and stars are alive). Societies with mean IQs no higher than those of the child duplicate the child's worldview, and history must be interpreted in these terms. Oesterdiekhoff believes that five evolutions produced modern society and go hand in hand: science, industrialism, enlightenment, humanist ethics, and democracy.

He cites Weber to the effect that the replacement of magic by scientific explanations was necessary to establish Western capitalism and that the enduring adherence of the Chinese to magic hindered their way to capitalism. Medieval people retained child-like beliefs like magic and religion, and tried animals in court to determine if they were guilty of crimes. Only after the scientific and industrial revolutions upgraded the minds of both the elite and the masses did genuine modern democracy become possible. Cognitive transformations dictated that the West became the birthplace of modern law and mores. As for humanist ethics, I have tried to show how habits of mind like using logic on abstractions and taking the hypothetical seriously tend to purify morality of racism and cruelty.

It is no accident that my first foray into the theory of intelligence cited Oesterdiekhoff as one of my inspirations: "I want to say that [he] brought a Piagetian interpretation of the past to my attention" (Flynn, 2007, p. 82).

Bridging theories

I. The Dickens/Flynn model

This model has been described in detail elsewhere (Flynn, 2009, 2012a) and acknowledged here. However, I promised to elaborate on it and will do so by focusing on the theme of how it liberated thinking about the causes of IQ trends over time. It did so by way of a theory that encompassed both individual differences and IQ trends between the generations.

Jensen noted that the twin studies showed that the impact of systematic environment was minimal, and that this was the environment, as distinct from chance environment, which would be likely to separate groups including races and generations: factors like SES, education, nutrition, and so forth. In other words, if environment was so weak, how could environmental changes from one generation to another cause huge score gains over time (after all, genetic differences from one generation to another would be minimal). Therefore, generational IQ gains must be suspect, perhaps the result of something like test sophistication. This assumes that the environmental factors that separate the generations are analogous to those that rank people for individual differences within a generation. But that seemed to be the case: more education pays off for one individual versus another, and more education would affect this generation as compared to the last.

Lewontin provided a "solution" that merely made the problem seem more intractable. He imagined a bag of seed that had plenty of genetic variation. It is randomly divided into two lots. One lot is grown in soil that is uniformly optimal; the other is grown in soil that is also uniform from one seed to another but the soil is missing a trace element of zinc. Here we have a situation where heritability within each lot of plants is 100 percent (this must be so since all environments are the same for each). However, the plants from the two lots of seed would differ in average height because one lot has an environment with no zinc. So while genes are totally dominant within each group (in accord with the twin studies), environment is the total cause of the between-group difference in average height.

As Jensen pointed out, this scenario assumes a mysterious Factor X. To get its potency, the between-group factor must be like the zinc. It must affect everyone within one group and be totally absent within the other. If it varied within groups, its potency would boost the environmental component of individual IQ variance far beyond what the twin studies allowed (almost zero potency). In other words, to solve the dilemma, a factor must *either* challenge the twin studies (absurd) *or* satisfy the conditions of Factor X (even more absurd). What factor could you imagine that did not vary both within and between the generations, certainly not more education, or better health, or more cognitively demanding leisure, or more cognitively demanding jobs? This froze speculation about the causes of IQ gains for twenty years, until the model provided the answer: it offered an individual multiplier that operates within groups, and a social multiplier that operates between groups.

The *individual multiplier* means that genes are dominant within a generation. As we have seen, before school, cognitive abilities differ between children, partially because they have different genes and partially because they belong to different families, with the latter often being more important at that age. But your genes are with you throughout life, and each environment is transitory to be swamped by the next current environment, which has little memory of previous ones. Thus, after the child goes to school, teachers and peers begin to swamp family in terms of current environment. Unlike parents, these people do not ignore genes in providing an environment for the child. The genetic advantage may be slight to begin with. But as genes co-opt quality of environment, their effects are greatly multiplied. At every age, ability level offers preferment for a better or worse environment; for example, the brighter child does more study, enters an honors steam, goes to university, and interacts with peers on his or her intellectual level. Either enhanced or diminished quality of environment becomes highly correlated with quality of genes (chance aside), and by adulthood genes dominate prediction of IQ. Thus, twin studies show that individuals are differentiated for ability primarily by genes, with systematic environment very weak as an independent factor.

Between generations, the *social multiplier* raises average cognitive performance often dramatically. Social change may offer (and demand) more schooling, better nutrition, more cognitively demanding work, and more cognitively demanding leisure. These environmental factors initially trigger a mild rise in average performance, but this becomes greatly magnified by feedback mechanisms. As the average performance rises, the rising mean itself becomes a powerful engine. Better average performance tends to enhance the performance of every individual, which raises the mean further, which further enhances the performance of each individual, and so forth. As parents see other parents keeping their children in school longer, they each tend to keep their children in school longer, and there is an education explosion from everyone with an average of six years of schooling, to everyone with a grade school education, to everyone with high school, to over half with some tertiary experience. As formal schooling frees people's minds from the concrete world to using logic on abstractions, performance on Raven's soars. And, since genes change little in one generation, the causal factors are almost entirely environmental.

Note that the social multiplier is not a Factor X. It does its job without affecting everybody equally. The educational revolution increased my years of school compared to my father by eight years (the difference between his eight years and my sixteen years for a Ph.D.). For most people, the generational difference would have been somewhat less, but this is irrelevant to the impact of the social multiplier.

Whether genes or environment are dominant depends on whose hand is on the throttle of a multiplier: genes use the individual multiplier to dominate individual differences within a generation (as measured by twin studies); environment uses the social multiplier to overpower genes from one generation to another (measured by IQ gains over time). We have solved the problem: we need *neither* ignore the results of the twin studies *nor* posit environmental factors between the generations that meet the absurd requirements of a Factor X. The environmental factors at work are pretty much the same within a generation and between generations. But the dynamics of how they operate are different, and it is this that explains how feeble environment can cause huge gain over time. Thus, our minds thawed and serious investigation of the causes of cognitive gains over time began.

II. van der Maas

Van der Mass et al. (2006), present a model endorsed by six colleagues at the University of Amsterdam. It shows that g need not be posited as some kind of underlying trait. I want to add that unless g is posited as such, it cannot qualify to play the role of an investment, or of an irreplaceable fuel that engenders conscious problem-solving. They show that g would arise automatically from a kind of reciprocal causation that undoubtedly takes place: the beneficial interaction between various cognitive processes during a child's development. The last chapter details my views that this is an important factor, but not the only factor, that creates g. But the fact that this factor can be modeled to show that it alone would engender g lends their model great significance.

For example, if you have better short-term memory (digit span), that helps you to hold in mind what you need to solve more complex problems (posed by the other Wechsler subtests), and better solutions make it possible to increase the efficiency of short-term memory. As we saw in Chapter 7, development of vocabulary allows you to vebalize Raven's problems (proves a help in logical analysis of abstractions), and developing your analytic powers can enhance your vocabulary (you can interact with better-educated people). Therefore, neither skill (neither Raven's as a measure of fluid g nor Vocabulary as a measure of crystallized g) "underlies" the other. This kind of beneficial causal interaction is not limited to one cognitive skill interacting with another. Performance which brings success will increase your motivation, which will in turn make for greater success. Abstract thinking may help to find creative solutions for interpersonal social or emotional problems, whereas good control over emotional and social life is beneficial to academic success.

Recall, it is the tendency of people who do above average on one Wechsler subtest to also do better than average on all the others that creates "the postive manifold" (the collection of positive correlations between all the subtests). And it is factor analysis of the manifold that yields g as an "underlying factor of general intelligence." But now we know that positing such an underlying factor is unnecessary: g can arise purely on the level of the skills the subtest measure, if they have beneficial interactions with one another. We simply do not need an underlying factor whose superiority makes us superior on all the Wechsler subtests; rather, superiority on each and every Wechsler subtest can make you superior on each and every one of the others.

They remark that their model is consistent with current explanations of massive IQ gains over time – that is, it could be expanded to do the job of the Dickens/Flynn model. This is true because the D/F model also uses mutually beneficial causal interactions: the individual multiplier emphasizes the interaction between an above average skill and above average environments; the social multiplier emphasizes the interaction between the rising cognitive quality of the social environment and the rising quality of each individual's environment. However, to explain what the D/F model explains, the expansion of the Dutch model would have to include facets that duplicate both the individual and social multipliers. Thus, it would be a complementary model and not a rival. As far as I can determine no true alternative to the D/F has been advanced and thus, by default, it is unique.

Brain physiology

By their very nature, theories of brain physiology link that level with individual differences. First, we want to map the areas/networks that are activated when people perform various cognitive skills; then, we will want to observe differences in those areas/networks that rank people's performance for each cognitive skill. In principle, brain physiology should also illuminate cognitive trends from one generation to another. It is a plausible hypothesis that as people began to drive motorcars, more mapping exercise enlarged the hippocampus between 1900 and today; and that the introduction of automatic guidance systems will erode the size of the hippocampus in the future. We must wait for data about the future but could project back into the past by studying drivers versus non-drivers – or ethnic groups that do not drive cars (such as the Amish).

As for group differences, there may well be a difference between the brains of blacks and whites. Take Jensen's evidence that as cognitive tasks become more complex, the black score deficit rises. Assume that I am correct: that blacks exercise their brains on complex cognitive problems less because their subculture demands that they solve such problems less often. Since the brain is like a muscle, then black brains on average would show less development of the areas/networks that deal with Raven's-type problems than white brains would. While there might be no difference for those brain areas/networks that "did" rote memory.

Of course, it may be that black genes limit growth of complex brain areas even if exercise is identical, or that a combination of "inferior" genes and less exercise is the cause. While this finding would be of great interest, it is difficult to see how brain imaging could solve the question of whether the black-white IQ gap is genetic or environmental because either hypothesis would account for the observations. I fear that magnetic resonance imaging of the brains will not be decisive and we must wait for direct genetic evidence: identification of the combination of genes that affect various kinds of cognition, and direct evidence that certain of their attributes lead to inferior performance, and direct evidence that these are more frequent in the black population. As for now, I stand by my analysis of black subculture as indicative that genetic equality is more probable than not.

Finally, all brain theories will have to resist the temptation of reductionism. However adequate our models of the brain become, physiology cannot replace psychology and sociology in the sense that we will still need multiple explanations on all three levels of human behavior. Physiology may be able to predict exactly who will be the best basketball player but we still need to know why someone is doing something as trivial as running around a court to try to throw a ball through a hoop, and why basketball became more popular after World War II, so that participation rose and triggered a huge rise in standards of performance by way of the social multiplier.

I. Ian Deary

Deary denies that he has a theory. However, Deary, Penke, and Johnson (2010) put forward concepts that signal an emerging theory on the level of brain physiology. Nerve cells or neurons actually carry on mental functions, such as analysis or information processing, and they are the gray matter of the brain (about 40 percent). Some nerve cells have fibrous extensions or axons. These are the communications network from one neuron to another (and between neurons and the other parts of the body), and they constitute the white matter of the brain (60 percent).

Crude correlations between IQ and greater volume of matter equal about 0.25 for the frontal cortex and the parietal and temporal cortices (all of which have been long thought to be the "seat" of intelligence), and for the hippocampus (spatial mapping). Correlations with gray matter have more to do with its thickness than its volume. Mental exercise enhances its thickness. It sprays it with dopamine, which thickens it, and makes them operate more efficiently the next time it is used; thus learning. White matter (axons) is sheathed in myelin, which is rather like wires being insulated. The myelin prevents the communications network between neurons from leaking electrical energy and losing efficiency.

There is a growing consensus that a "small-world network" is the optimum for communications. Preferable is a high level of local clustering of neurons and short pathways for the axions that link them – that is, the most efficient network is few paths and short paths to link the clusters. The best brains process information more efficiently because they use fewer brain resources to do reasoning tasks. The average brain has to be very active to deal with a mental task of moderate difficulty, while a superior brain solves it with less effort. For difficult tasks, the average brain is inactive because it gives up, while the superior brain now mobilizes all of its resources. With age, neurons lose plasticity, and dopamine has less effect in repairing them. Recall that myelin insulates the axons and enhances efficiency. Myelin breakdown and repair is continually occurring over the brain's entire neural network, but in old age we begin losing the repair battle.

II. Roberto Colum

Most working in the area of brain research accept the above model but some make their own important contributions. Barbey et al. (2014) studied human brain lesions. They estimate the degree to which individual differences in cognitive performance (WAIS scores, measured emotional intelligence, and personality inventories) are predicted by psychological variables; and then use lesion mapping to lay bare an underlying shared network of frontal, temporal, and parietal brain regions (including tracks that bind these areas into a coordinated system). Colum et al. (2012) cite further evidence in favor of the "standard" model. Even brief mental exercise enhances both IQ and cortical thickness (gray matter that acts as a communications network). In addition, subjects that suffer a quick and sharp IQ decline display significant reductions in cortical thickness. This tallies with the Dickens/Flynn model: current cognitive environment quickly swamps past environments.

There is the tantalizing possibility that direct stimulation of the brain can influence its development. Colum cites Santarnecchi et al. (2013): gamma-band stimulation over the left middle frontal gyrus enhanced fluid intelligence performance. The participants needed less time to solve complex problems from Raven's Advanced Progressive Matrices test. He notes two new projects likely to enhance our understanding through new technology, and I now turn to them.

III. BRAIN Initiative

The US National Institute of Health (2014) has launched a ten-year project to advance brain research. Advances in electrical, optical, acoustic, and genetic techniques will inform us about molecules, cells, circuits, systems, and cognitive behavior. Technologies would include implantable devices with combined recording and stimulation capabilities. For the first time, we will have a dynamic picture of the brain that shows how individual cells and complex neural circuits interact in both time and space at the speed of thought.

In particular, the project will seek:

- the neural circuits that underlie the ability to represent information symbolically (as in language) and use that information in novel situations;
- the neural circuits that enable mental mathematical calculations; and
- the patterns of neural activity that correspond to human emotional states.

IV. The Human Brain Project

The Human Brain Project (2014) has put its manifesto online, another ten-year project this time funded by the European Union.

Its objective seems a bit more modest: a first draft of rodent and human brain models. The new technologies include novel super-computing software and hardware, analysis software, algorithms, search technology, and much more. These sound less pictoral and more computational than the US technologies. However, its ultimate aim is similar: modeling the brain as a multi-level integrated system from genes all the way to cognition.

Answers

I hope the diversity of theories, and the brave new world brain research promises, have not distracted you from the purpose of this chapter, which is to answer certain questions about the compatibility of various theories of intelligence.

(1) Meta-theory and scientific theories: insofar as theories do not move a theory-embedded concept up to the level of meta-theory, and do not take a concept viable in one area and apply it to another (like Jensen), all scientific theories are compatible with my meta-theory.

(2) Compatibility of scientific theories with one another:

Individual differences: theories compete with one another. Some have put forward hypotheses that have been falsified; for example, the investment hypothesis (*g* theories). Others thus far have escaped falsification (Sternberg). There is considerable debate about what skills are socially important and what skills have a significant cognitive content (Gardner). Theories differ in terms of the range of non-cognitive personal traits that need to be identified (PPIK, Bandura, PASS).

Differences between generations and other groups: I believe Flynn's theory and the Dickens/Flynn model have explanatory power and await genuine alternatives. Oesterdiekhoff's theory is not an alternative but compatible, at least insofar as Raven's and Similarities gains are concerned.

Brain physiology: I see no incompatibility between various theories at present and anticipate significant advances over the next decade.

11   Psychology and Cardinal Bellarmine

What I have to say at the end is worth only a page. The science of psychology shows that those who are pre-scientific in their worldview must think again. They must all recognize that science is the best instrument to explore the real world, including the real world of human behavior, and that "common sense" cannot compete. However, I sympathize with Cardinal Bellarmine in his advice to Galileo: science should revise scripture (what ordinary people think about the significance of what they do) only when the evidence is decisive and interpretation of its consequences lucid.

Like Pinker (2002), I have no sympathy with those who believe that human nature is a blank slate that environment may do with as it will. I accept the main thrust of the twin studies. But I reject post-twin pessimism. I also have little sympathy with those who interpret the twins as a genetic veto on our sense of social justice and our efforts to improve our children, our selves, and our species. Whether this short book represents what the best science has to say, and whether it represents what most people suspect to be true, and whether I have reconciled the two, readers both expert and general can decide at leisure.

I have struck another note of optimism. The distinction between the meta-theory of psychology and scientific theories of psychology should prevent us from repeating the mistakes of the past. In addition, the meta-theory's heuristics offers good advice, good enough to allow scientific theories to get on with their job. These theories compete with one another in terms of explanation and prediction, and it will be sad if all of them are not eventually transcended. But I do not foresee the kind of failure that would require the radical step of a new heuristic.

For scholars who wish to use the Age-Table Method to measure family effects in nations other than the USA

Wechsler

Publisher	Test	Language
Pearson Australia	WAIS-R	English (Aus)
Psykologien Kustannus OY	WAIS-R	Finnish
ECPA (Les Éditions du Centre de Psychologie Appliquée) (Pearson France sas)	WAIS-R	French (France)
Giunti OS Organizzazioni Speciali	WAIS-R	Italian
Nihon Bunka Kagakusha Co. Ltd.	WAIS-R	Japanese
Pracownia Testów Psychologicznych (PTP)	WAIS-R	Polish
Chinese Behavioral Science Corporation	WAIS-III	Chinese
Pearson Assessment and Information AB	WAIS-III	Danish
Pearson Assessment & Information BV	WAIS-III	Dutch
Pearson Australia	WAIS-III	English (Aus)
Pearson Canada Assessment	WAIS-III	English (Canada)
Pearson Assessment UK	WAIS-III	English (UK)
Psykologien Kustannus OY	WAIS-III	Finnish
Pearson Canada Assessment	WAIS-III	French (Canada)
ECPA (Les Éditions du Centre de Psychologie Appliquée) (Pearson France sas)	WAIS-III	French (France)
Pearson Assessment and Information GmbH	WAIS-III	German
PsychTech Ltd.	WAIS-III	Hebrew
Icelandic Psychological Measures	WAIS-III	Icelandic
Nihon Bunka Kagakusha Co. Ltd.	WAIS-III	Japanese

(continued)

Publisher	Test	Language
Vilnius University, Laboratory of Special Psychology	WAIS-III	Lithuanian
Pearson Assessment and Information AB	WAIS-III	Norwegian
Casapsi Livraria e Editora Ltda	WAIS-III	Portuguese (Bz)
Editorial Paidos, SA	WAIS-III	Spanish (Argentina)
Editorial el Manual Moderno SA de CV	WAIS-III	Spanish (Mexico)
Pearson Assessment and Information AB	WAIS-III	Swedish
Beijing Healthmen Company	WAIS-IV	Chinese
King-May Psychological Assessment	WAIS-IV	Chinese
Pearson Assessment and Information AB	WAIS-IV	Danish
Pearson Assessment & Information BV	WAIS-IV	Dutch
Jopie van Rooyen & Partners SA (Pty) Ltd.	WAIS-IV	English (adapted)
Pearson Australia	WAIS-IV	English (Aus)
Pearson Canada Assessment	WAIS-IV	English (Canada)
Pearson Assessment UK	WAIS-IV	English (UK)
Psykologien Kustannus OY	WAIS-IV	Finnish
Pearson Canada Assessment	WAIS-IV	French (Canada)
ECPA (Les Éditions du Centre de Psychologie Appliquée) (Pearson France sas)	WAIS-IV	French (France)
Pearson Assessment and Information GmbH	WAIS-IV	German
Motibo Publishing SA	WAIS-IV	Greek
Giunti OS Organizzazioni Speciali	WAIS-IV	Hungarian
Giunti OS Organizzazioni Speciali	WAIS-IV	Italian
Korea Psychology Co.	WAIS-IV	Korean
Pearson Assessment and Information AB	WAIS-IV	Norwegian
Universidad Catolica de Chile	WAIS-IV	Spanish (Chile)

Publisher	Test	Language
Editorial el Manual Moderno SA de CV	WAIS-IV	Spanish (Mexico)
Pearson Educacion, SA (Espana)	WAIS-IV	Spanish (Spain)
Pearson Assessment and Information AB	WAIS-IV	Swedish

Note: the above is a list of Wechsler adaptations that have been licensed out and published. There are many recent Wechsler editions that have been licensed out but that have not yet been published. There is no guarantee that all of the above have adapted manuals; this is left up to the local publishers, who decide which components of the test they want to / need to translate for their local markets. Although the list refers only to the WAIS, in virtually every case the WISC (and sometimes the WPPSI) is available as well.

Stanford-Binet

Many of the nations listed have not normed the test on their own standardization sample and, thus, do not provide the necessary tables to convert raw scores into standard scores by subtest and by age. Australia, Canada, and Ireland may be exceptions. There are translations in progress in Germany and Poland. Only the manuals of the SB-4 and SB-5 are usable.

For contacts in various nations, email Elizabeth Allen eallen@proedinc.com:

- Australia
- Canada
- Great Britain
- Hong Kong
- India
- Ireland
- Israel
- Malaysia
- Thailand

Appendix I: Wechsler Vocabulary and description of method of analysis

Sources

Wechsler, D. (1949). *Wechsler Intelligence Scale for Children: Manual.* New York: The Psychological Corporation. **WISC data**

Wechsler, D. (1955). *Wechsler Adult Intelligence Scale: Manual.* New York: The Psychological Corporation. **WAIS data**

Wechsler, D. (1974). *Wechsler Intelligence Scale for Children: Revised.* New York: The Psychological Corporation. **WISC-R data**

Wechsler, D. (1981). *Wechsler Adult Intelligence Scale: Revised.* New York: The Psychological Corporation. **WAIS-R data**

Wechsler, D. (1989). *Wechsler Preschool and Primary Scale of Intelligence: Revised.* San Antonio, TX: The Psychological Corporation. **WPPSI-R data**

Wechsler, D. (1992). *Wechsler Intelligence Scale for Children – Third Edition: Manual (Australian Adaptation).* San Antonio, TX: The Psychological Corporation. **WISC-III data**

Wechsler, D. (1997). *Wechsler Adult Intelligence Scale – Third Edition: Manual.* San Antonio, TX: Pearson. **WAIS-III data**

Wechsler, D. (2002). *Wechsler Preschool and Primary Scale of Intelligence – Third Edition: Manual.* San Antonio, TX: Pearson. **WPPSI-III data**

Wechsler, D. (2003). *Wechsler Intelligence Scale for Children – Fourth Edition: Manual.* San Antonio, TX: The Psychological Corporation. **WISC-IV data**

Wechsler, D. (2008). *Wechsler Adult Intelligence Scale – Fourth Edition: Manual.* San Antonio, TX: Pearson. **WAIS-IV data**

The analysis of this subtest will be prefaced by a step-by-step description of how the method works. The method takes you from the data presented in the age tables (for a particular subtest) to conclusions about the extent to which family or common environment influences performance on whatever ability the subtest measures, and at what age (if any) that influence disappears. Since the method is fundamentally the same throughout, it will not be repeated for other subtests; which is to say, all other prefatory remarks will refer you back to this description.

The Age-Table Method: its assumptions

(1) Those whose performance puts them at +2 SD above the median come from homes whose cognitive quality will extend far below that level. As long as family environment persists, they will be at a disadvantage when compared to whatever age the lingering effects of family environment fade in favor of a match between genes and current environment (where it becomes a null factor). The same is true, although to a lesser degree, for those at +1 SD above the median.

(2) Those at −2 SD below the median will come from homes whose cognitive quality will extend far above that level. As long as family environment persists, they will be at an advantage when compared to whatever age this disappears (where it is a null factor). The same is true for those at −1 SD below the median.

(3) At all ages, those at the median will come from as many homes above as below the average level of cognitive quality. Therefore, they can serve as a criterion against which the disadvantages/advantages of high and low performers can be measured.

(4) To estimate the size of these disadvantages/advantages, it is necessary to compare them to an age at which family

environment is presumed to disappear. This is usually a matter of norming the performance at a given age (at all levels) on that "target" age, although sometimes (as here) the reverse makes sense. I put the target age at that at which performances peak – before age erodes performance. On some subtests, this age is too young for "old age" to weigh in, but it is the best we can do. In data where it is simply a matter that the age tables do not extend beyond youth, such as the 1985 Vocabulary subtest of the Stanford-Binet, comparative data can make good the lack.

The Age-Table Method dictates: four tasks

I. Enter all the relevant data for all Wechsler subtests. A choice is sometimes necessary at the bottom of the scale, which may assign 3 SD below the median either nil or a wide range of raw scores. When this clearly inflates the raw score value of the bottom SD (between −2 and −3 SD), I was guided by the SD above (the raw score difference between −1 and −2 SD). This guards against an inflated estimate of the influence of family environment at the lowest level.

II. Compare the raw score at a given age at the +2 SD level to the raw score of the target age at the +2 SD level. Take the raw score difference (between the two) expressed in SDs times 15 to convert it into IQ points. Do this for all levels including the median. The difference at the median (see text) is a pure measure of enhanced performance with maturity, and must be subtracted at all levels – thus the next task.

III. Subtract the difference at the median from the difference at all other levels. If family environment is potent, this must lead to an asymmetry: the +2 and +1 SD levels will give a plus, and the −1 and −2 SD levels will give a minus.

IV. To estimate the proportion of variance family environment accounts for, there must be a criterion of the cognitive value of the home and a measure of how well that correlates with performance. This gives a correlation at each of the four levels and these are averaged. The correlation squared gives the proportion of variance explained. For the way in which the cognitive quality of homes at the various levels is estimated, see the text: for example, it is assumed that those who perform at +2 SD do not come from homes in the bottom 30 percent of quality.

The sections to follow will show how these tasks are to be performed by spelling out the computations.

Step I

Wechsler vocabulary data: raw scores by age at various performance levels. Values in bold are emphasized as key in the computations that follow.

WAIS						
	16–17	18–19	20–24	25–34	**35–44**	45–54
+3 SD	75.5					
+2 SD	64	67	69	**72**	**74**	74.5
+1 SD	49.5	55	58.5	62	**65.5**	63.5
Med.	31	37	41	44	**43.5**	42
−1 SD	19.5	22.5	22.5	24.5	**25.5**	24
−2 SD	11	11	11	14	**12**	12

WAIS-R					
	16–17	18–19	20–24	**25–34**	35–44
+3 SD	68				
+2 SD	61	61.5	65	**67**	66.5
+1 SD	50.5	52	57.5	**62**	59.5
Med.	36.5	40	46	**50.5**	49.5
−1 SD	18.5	19.5	30	**35**	29.5
−2 SD	9	9.5	12	**12**	12	5

WAIS-III										
	16–17	18.0	18–19	20–24	25–29	30.0	30–34	35–44	**45–54**	55–64
+3 SD	64									
+2 SD	54	(55.25)	56.5	57.5	60	(60.25)	60.5	62.5	**62.5**	62.5
+1 SD	45	(47)	49	49	51	(52)	53	55	**56**	53.5
Med.	33.5	(34.75)	36	36	40	(41.25)	42.5	44.5	**46.5**	41
−1 SD	21.5	(22.5)	23.5	23.5	26.5	(28.25)	30	**32**	26	
−2 SD	11	(11.25)	11.5	14	14.5	(15)	15.5	16.5	**17**	13.5

WAIS-IV										
	16–17	18–19	20–24	25–29	30.0	30–34	35–44	**45–54**	**55–64**	65–69
+3 SD	55									
+2 SD	47.5	49.5	50.5	52	52.5	53	54	**54**	54	54
+1 SD	40	41	43	45	46	47	48.5	**49**	49	48
Med.	30	31	33	34.5	35.25	36	37.5	**38.5**	38.5	37.5
−1 SD	19.5	20.5	22.5	23.5	24	24.5	25.5	**25.5**	25.5	25
−2 SD	7.5	8.5	10.5	11.5	12	12.5	13.5	**13.5**	13.5	12

WISC					
7 = tables 6/8–6/11; 9.5 = 9/4–9/7; 12 = 11/8–11/11; 14.5 = 14/4–14/7; 16 = 15/8–15/11					
	7	9.5	**12**	14.5	16 (17 in the master table to follow)
+2 SD	31	44	54	62.5	65.5
+1 SD	25.5	36	45	54	57
Med.	19.5	29	38	46	47.5
−1 SD	13.5	22	29	36	38
−2 SD	7.5	13.5	19	24.5	25.5
−3 SD			11		17.5

WISC-R

7 = tables 6/8–6/11; 9.5 = 9/4–9/7; 12 = 11/8–11/11; 14.5 = 14/4–14/7; 17 = 16/8–16/11

	7	9.5	**12**	14.5	17
+2 SD	26	39.5	50	57.5	60
+1 SD	21.5	32.5	43.5	51	55.5
Med.	17	26.5	35	42	48
−1 SD	12	20.5	27	32	35.5
−2 SD	8	14.5	20.5	23.5	26
−3 SD			(13.5)		(16.5)

WISC-III

7 = tables 6/8–6/11; 9.5 = 9/4–9/7; 12 = 11/8–11/11; 14.5 = 14/4–14/7; 17 = 16/8–16/11

	7	9.5	**12**	14.5	17
+2 SD	23.5	36.5	46.5	53	57
+1 SD	19	30.5	39.5	47.5	52.5
Med.	14.5	24.5	31	40	44
−1 SD	10	18.5	24.5	29	33
−2 SD	7	13.5	18.5	22.5	26.5
−3 SD			(12.5)		(20)

WISC-IV

7 = tables 6/8–6/11; 9.5 = 9/4–9/7; 12 = 11/8–11/11; 14.5 = 14/4–14/7; 17 = 16/8–16/11

	7	9.5	**12**	14.5	17
+2 SD	32	45.5	52.5	59.5	63.5
+1 SD	25.5	38	44.5	52	56
Med.	19.5	30.5	36.5	43	47
−1 SD	13.5	23	28	33	39
−2 SD	7.5	15	20	25	30
−3 SD			(12)		(21)

WPPSI-R			
$3 = 2/11.5 – 3/2.5$; $4 = 3/11.5 – 4/2.5$; $7 (6.75) = 6/8.5 – 6/11.5$			
	3	4	7
+2 SD	18	24	38
+1 SD	13.5	18.5	34
Med.	8.5	13.5	27.5
−1 SD	4	7.5	19
−2 SD	1	3	14.5
−3 SD		—	(9)

WPPSI-III		
$4 = 4/0 – 4/2$; $7 = 6/8 – 6/11$. Note: although, below 4, there is a "receptive vocabulary" test (recognizing words), its raw scores are not comparable to later years.		
	4	7
+2 SD	24	38.5
+1 SD	19.5	33.5
Med.	13.5	26
−1 SD	7.5	18
−2 SD	3	9.5
−3 SD		(1)

Step II

How to get the values by level (+2 to −2 SD).

Plan: for adult ages (17 and over)

(1) Age 17 is almost duplicated in both the WISC and WAIS data.

(i) WAIS has 16–17 data but these run from 16.0 to 17.11, so 17 is the mean.

(ii) WISC has 16.8 to 16.11, so 16.83 years is mean, close enough – the WISC itself ends a year earlier, so there I had to make do with 15.83 years.

(2) To link the WISC and WAIS data for those aged 12 (really 11.83 years) and aged 14.5 (exact; used 14.4–14.7) I used age 17 as the link.

   (i) For example, age 12 was normed on WISC 17 at levels from +2 to −2 SD.

   (ii) The adult maximum was normed on WAIS 17.

   (iii) That result was added to the WISC result to get a comparison between age 12 and ages 35–44. For example, see first row below in Table AI1A: 19.74 (12) + 13.04 (17) = 32.78 (12 and adult age compared).

   (iv) For ages 17 and above, the result is simply a matter of the difference between that age and the adult maximum (target age), which is then normed on WAIS age 17.

An example should make this clear – take the value for 12 in the first row:

	(1) WAIS data									
	16–17	18–19	20–24	25–34	**35–44**	45–54	55–64	65–69	70–74	75+
+3 SD	**75.5**									
+2 SD	**64**	67	69	72	**74**	74.5	74.5	72	70.5	70

74 − 64 = 10; 75.5 − 64 = 11.5 (SD); 10 ÷ 11.5 = 0.870 SD difference; 0.870 × 15 = **13.04** IQ difference

	(2) WISC data		
	12	14.5	16 (year 17 in table)
+2 SD	**54**	62.5	**65.5**
+1 SD	45	54	**57**
Med.	38	46	**47.5**

65.5 − 54 = 11.5; SD, age 17, between +2 and +1 = 8.5; so difference is 1 SD with 3 raw score point left; SD, age 17, between +1 and median = 9.5; 3 ÷ 9.5 = 0.32 SD; 1 + 0.32 = 1.32 total SD difference; 1.32 × 15 = **19.74** IQ difference.

(3) So 13.04 + 19.74 = **32.78** as the total difference (at the +2 SD level) for age 12 compared to ages 35–44, using age 17 as the link.

Once you have a value comparing age 12 and the adult maximum you can use that age as a link by norming earlier ages (ages 7 and 9.5) on age 12. For example, in the first row, add 41.67 to 32.78 and you get 74.45 for age 7 compared to ages 35–44.

All values below (Table AI1A) were calculated in this way.

Once you have a value comparing age **7** and the adult maximum, you can use that age as a link by norming earlier ages (using the WPPSI data) on age 7. For example, in the first row, add 50.03 to 85.84 and you get 135.87 for age 3 compared to ages 45–54.

I then decided to add ages 25–35 for both all data and recent data (WAIS-IV only: marked as such). The calculation of the difference of high and low values from the median is anticipated here.

+2 SD (1950.5)	2.61 − −0.41	= +3.02
+2 SD (1975)	nil	
+2 SD (1992)	3.38 − 6.85	= −3.47
+2 SD (**2004.5**)	3.00 − 4.88	= −**1.88** (**recent data**)
Average ÷ 3		−**0.78** (**all data**)
+1 SD (1950.5)	3.62 − −0.41	= +4.03
+1 SD (1975)	nil	
+1 SD (1992)	5.45 − 6.85	= −**1.40**
+1 SD (**2004.5**)	6.00 − 4.88	= +**1.12** (**recent data**)
Average ÷ 3		+**1.25** (**all data**)
Median (1950.5)	−0.41 − −0.41	= nil
Median (1975)	nil	
Median (1992)	6.85 − 6.85	= nil
Median (**2004.5**)	4.88 − 4.88	= **nil** (**recent data**)
Average ÷ 3		**nil** (**all data**)
−1 SD (1950.5)	1.30 − −0.41	= +1.71
−1 SD (1975)	nil	
−1 SD (1992)	5.94 − 6.85	= −0.91
−1 SD (**2004.5**)	2.14 − 4.88	= −**2.74** (**recent data**)
Average ÷ 3		−**0.65** (**all data**)
−2 SD (1950.5)	−3.53 − −0.41	= − 3.12
−2 SD (1975)	nil	
−2 SD (1992)	2.86 − 6.85	= −3.99
−2 SD (**2004.5**)	1.88 − 4.88	= −**3.00** (**recent data**)
Average ÷ 3		−**3.37** (**all data**)

Table A11A Adult versus child IQ differences at four times at five IQ levels (all normed on 17-year-old curve)

Child age	Compared to adults			Compared to adults								Adults
	17	18–19	20–24	To age 12				To age 17				
				7	9.5	12	14.5	7	9.5	12	14.5	
+2 SD (1950.5)	13.04	9.13	6.52	41.67	17.14	19.74	5.29	74.45	49.92	32.78	18.33	35–44
+2 SD (1975)	12.86	11.79	4.29	47.31	22.06	26.00	8.33	86.17	60.92	38.86	21.19	25–34
+2 SD (1992)	12.75	9.00	7.50	47.50	20.29	25.59	13.33	85.84	58.63	38.34	26.08	45–54
+2 SD (**2004.5**)	13.00	9.00	7.00	37.94	13.13	20.83	8.00	71.77	46.96	33.83	21.00	45–64
Average	**12.91**	**9.73**	**6.33**					**79.56**	**54.11**	**35.95**	**21.65**	
+1 SD (1950.5)	16.55	10.86	7.24	35.25	18.33	18.95	4.74	70.75	53.83	35.50	21.29	35–44
+1 SD (1975)	16.43	14.29	6.23	42.69	19.69	20.04	9.00	79.16	56.16	36.47	25.43	25–34
+1 SD (1992)	18.33	11.67	11.67	43.75	16.15	21.14	8.82	83.22	55.62	39.47	27.15	45–54
+1 SD (**2004.5**)	18.00	16.00	12.00	34.69	12.19	19.69	6.67	72.38	49.99	37.69	24.67	45–64
Average	**17.33**	**13.21**	**9.29**					**76.38**	**53.90**	**37.28**	**24.64**	
Med. (1950.5)	10.14	5.27	2.03	29.25	15.00	15.00	2.37	54.39	40.14	25.14	12.51	35–44
Med. (1975)	15.00	11.25	4.82	37.50	16.15	15.79	7.20	68.29	46.94	30.79	22.20	25–34
Med. (1992)	16.96	13.70	13.70	40.00	15.00	19.62	5.45	76.58	51.58	36.58	22.41	45–54
Med. (**2004.5**)	12.75	11.25	8.25	30.94	10.59	19.17	7.50	62.86	42.51	31.92	20.25	45–64
Average	**13.71**	**10.37**	**7.20**					**65.53**	**45.29**	**31.11**		
−1 SD (1950.5)	7.83	3.91	3.91	25.31	10.50	10.80	2.40	43.94	29.13	18.63	10.23	35–44
−1 SD (1975)	13.75	12.92	4.17	32.81	12.86	13.42	5.53	59.98	40.03	27.17	19.23	25–34
−1 SD (1992)	13.13	10.63	10.63	36.25	15.00	19.62	9.23	69.00	47.75	32.75	22.36	45–54
−1 SD (**2004.5**)	8.57	7.14	4.29	27.19	9.38	18.33	10.00	54.09	36.28	26.90	18.57	45–64
Average	**10.82**	**8.65**	**5.75**					**56.75**	**38.30**	**26.36**	**17.60**	
−2 SD (1950.5)	1.76	1.76	1.76	21.56	10.31	12.19	1.88	35.51	24.26	13.95	3.64	35–44
−2 SD (1975)	4.74	3.95	0.00	26.79	12.86	8.68	3.95	40.21	26.18	13.42	8.69	25–34
−2 SD (1992)	8.57	7.86	4.29	28.75	12.50	18.46	9.23	55.78	39.53	27.03	17.80	45–54
−2 SD (**2004.5**)	7.50	6.25	3.75	23.44	9.38	16.67	8.33	47.61	33.55	24.17	15.83	45–64
Average	**5.64**	**4.96**	**2.45**					**44.78**	**30.88**	**19.64**	**11.49**	

Table AI1B Adult versus (young) child IQ differences at two times at five IQ levels (all normed on 17-year-old curve)

Compared to:				Adults			
Normed against:	Age 4	Age 7	Age 7				Adult age
Child age	3	4	3	7	3	4	
+2 SD (1992)	13.85	36.18	**50.03**	**85.84**	**135.87**	122.02	45–54
+2 SD (2004.5)	—	33.75	—	71.77	—	105.52	45–64
Average						**113.77**	
+1 SD (1992)	15.00	31.67	46.67	83.22	129.89	114.89	45–54
+1 SD (2004.5)	—	27.19	—	72.38	—	99.57	45–64
Average						**107.23**	
Med. (1992)	12.50	32.73	45.23	76.58	121.81	109.31	45–54
Med. (2004.5)	—	22.94	—	62.86	—	85.80	45–64
Average						**97.56**	
−1 SD (1992)	11.67	34.09	45.76	69.00	114.76	103.09	45–54
−1 SD (2004.5)	—	18.53	—	54.09	—	72.62	45–64
Average				**61.55**		**87.86**	
−2 SD (1992)	10.00	31.36	41.36	55.78	97.14	87.14	45–54
−2 SD (2004.5)	—	11.47		47.61		59.08	45–64
Average						**73.11**	

I then decided to carry Vocabulary through to older ages for recent data only (which = WAIS-IV). The calculation of difference of high and low values from the median is anticipated here.

25–29	+2 SD	$4.00 - 6.00 = -2.00$
	+1 SD	$8.00 - 6.00 = +2.00$
	Median =	$6.00 - 6.00 = -$
	−1 SD	$2.86 - 6.00 = -3.14$
	−2 SD	$2.50 - 6.00 = -3.50$
30–34	+2 SD	$2.00 - 3.75 = -1.75$
	+1 SD	$4.00 - 3.75 = +0.25$
	Median =	$3.50 - 3.50 = -$
	−1 SD	$1.18 - 3.50 = -2.32$
	−2 SD	$1.00 - 3.50 = -2.50$

35-44	+2 SD	$0.00 - 1.50 = -1.50$
	+1 SD	$1.00 - 1.50 = -0.50$
	Median =	$1.50 - 1.50 = -$
	−1 SD	$0.00 - 1.50 = -1.50$
	−2 SD	$0.00 - 1.50 = -1.50$

Step III

Subtract the difference at the median from the difference at all other levels. This calculation has been anticipated for older age groups.

Table AI2 Adult versus youth Vocabulary gaps: how much do gaps at levels above/below median differ from those at median? ALL data

Child age 7				
79.56 (+2 SD)	minus	65.53 (median)	equals	+14.03
76.38 (+1 SD)	minus	65.53 (median)	equals	+10.85
65.53 (median)	minus	65.53 (median)	equals	—
56.75 (−1 SD)	minus	65.53 (median)	equals	−8.78
44.78 (−2 SD)	minus	65.53 (median)	equals	−20.75
Child age 9.5				
54.11 (+2 SD)	minus	45.29 (median)	equals	+8.82
53.90 (+1 SD)	minus	45.29 (median)	equals	+8.61
45.29 (median)	minus	45.29 (median)	equals	—
38.30 (−1 SD)	minus	45.29 (median)	equals	−6.99
30.88 (−2 SD)	minus	45.29 (median)	equals	−14.41
Child age 12				
35.95 (+2 SD)	minus	31.11 (median)	equals	+4.84
37.28 (+1 SD)	minus	31.11 (median)	equals	+6.17
31.11 (median)	minus	31.11 (median)	equals	—
26.36 (−1 SD)	minus	31.11 (median)	equals	−4.85
19.64 (−2 SD)	minus	31.11 (median)	equals	−11.47

(continued)

Table AI2 Adult versus youth Vocabulary gaps: how much do gaps at levels above/below median differ from those at median? ALL data (*continued*)

		Child age 14.5		
21.65 (+2 SD)	minus	19.34 (median)	equals	+2.31
24.64 (+1 SD)	minus	19.34 (median)	equals	+5.30
19.34 (median)	minus	19.34 (median)	equals	—
17.60 (−1 SD)	minus	19.34 (median)	equals	−1.74
11.49 (−2 SD)	minus	19.34 (median)	equals	−7.85
		Child age 17		
12.91 (+2 SD)	minus	13.71 (median)	equals	−0.80
17.33 (+1 SD)	minus	13.71 (median)	equals	+3.62
13.71 (median)	minus	13.71 (median)	equals	—
10.82 (−1 SD)	minus	13.71 (median)	equals	−2.89
5.64 (−2 SD)	minus	13.71 (median)	equals	−8.07
		Ages 18–19		
9.73 (+2 SD)	minus	10.37 (median)	equals	−0.64
13.21 (+1 SD)	minus	10.37 (median)	equals	+2.84
10.37 (median)	minus	10.37 (median)	equals	—
8.65 (−1 SD)	minus	10.37 (median)	equals	−1.72
4.96 (−2 SD)	minus	10.37 (median)	equals	−5.41
		Ages 20–24		
6.33 (+2 SD)	minus	7.20 (median)	equals	−0.87
9.29 (+1 SD)	minus	7.20 (median)	equals	+2.09
7.20 (median)	minus	7.20 (median)	equals	—
5.75 (−1 SD)	minus	7.20 (median)	equals	−1.45
2.45 (−2 SD)	minus	7.20 (median)	equals	−4.75
		Ages 25–34 (see above)		
			equals	−0.78
			equals	+1.25
			equals	—
			equals	−0.65
			equals	−3.37

	Latest data		
Child age 7			
71.77 (+2 SD)	minus	62.86 (median)	equals	+8.91
72.38 (+1 SD)	minus	62.86 (median)	equals	+9.52
62.86 (median)	minus	62.86 (median)	equals	—
54.09 (−1 SD)	minus	62.86 (median)	equals	−8.77
47.61 (−2 SD)	minus	62.86 (median)	equals	−15.25
Child 9.5			
46.96 (+2 SD)	minus	42.51 (median)	equals	+4.45
49.99 (+1 SD)	minus	42.51 (median)	equals	+7.48
42.51 (median)	minus	42.51 (median)	equals	—
36.28 (−1 SD)	minus	42.51 (median)	equals	−6.23
33.55 (−2 SD)	minus	42.51 (median)	equals	−8.96
Child age 12			
33.83 (+2 SD)	minus	31.92 (median)	equals	+1.91
37.69 (+1 SD)	minus	31.92 (median)	equals	+5.77
31.92 (median)	minus	31.92 (median)	equals	—
26.90 (−1 SD)	minus	31.92 (median)	equals	−5.02
24.17 (−2 SD)	minus	31.92 (median)	equals	−7.75
Child age 14.5			
21.00 (+2 SD)	minus	20.25 (median)	equals	+0.75
24.67 (+1 SD)	minus	20.25 (median)	equals	+4.42
20.25 (median)	minus	20.25 (median)	equals	—
18.57 (−1 SD)	minus	20.25 (median)	equals	−1.68
15.83 (−2 SD)	minus	20.25 (median)	equals	−4.42
Child age 17			
13.00 (+2 SD)	minus	12.75 (median)	equals	+0.25
18.00 (+1 SD)	minus	12.75 (median)	equals	+5.25
12.75 (median)	minus	12.75 (median)	equals	—
8.57 (−1 SD)	minus	12.75 (median)	equals	−4.18
7.50 (−2 SD)	minus	12.75 (median)	equals	−5.25
Ages 18–19			
9.00 (+2 SD)	minus	11.25 (median)	equals	−2.25
16.00 (+1 SD)	minus	11.25 (median)	equals	+4.75
11.25 (median)	minus	11.25 (median)	equals	—
7.14 (−1 SD)	minus	11.25 (median)	equals	−4.11
6.25 (−2 SD)	minus	11.25 (median)	equals	−5.00

(*continued*)

Table AI2 Adult versus youth Vocabulary gaps: how much do gaps at levels above/below median differ from those at median? ALL data (*continued*)

		Ages 20–24		
7.00 (+2 SD)	minus	8.25 (median)	equals	*−1.25*
12.00 (+1 SD)	minus	8.25 (median)	equals	+3.75
8.25 (median)	minus	8.25 (median)	equals	—
4.29 (−1 SD)	minus	8.25 (median)	equals	−3.96
3.75 (−2 SD)	minus	8.25 (median)	equals	−4.50
		Ages 25–34 (see above)		
			Equals	*−1.88*
			Equals	+1.12
			Equals	—
			Equals	−2.74
			Equals	−3.00
		Ages 25–29 (see above)		
			Equals	*−2.00*
			Equals	+2.00
			Equals	—
			Equals	−3.14
			Equals	−3.50
		Ages 30–34 (see above)		
			Equals	*−1.75*
			Equals	+0.25
			Equals	—
			Equals	−2.32
			Equals	−2.50
		Ages 35–44 (see above)		
			Equals	*−1.50*
			Equals	−0.50
			Equals	—
			Equals	−1.50
			Equals	−1.50

The young children now get special tables because they involve only two times (1992 and 2004.5) when calculating the results of the total data. The latest data is, of course, merely the 2004.5 set.

Young children: how much do gaps at levels above/below median differ from those at median? ALL data

Child age 3 (only for 1992)				
135.87 (+2 SD)	minus	121.81 (median)	equals	+14.06
129.89 (+1 SD)	minus	121.81 (median)	equals	+8.08
121.81 (median)	minus	121.81 (median)	equals	—
114.76 (−1 SD)	minus	121.81 (median)	equals	−7.05
97.14 (−2 SD)	minus	121.81 (median)	equals	−24.67
Child age 4				
113.77 (+2 SD)	minus	97.56 (median)	equals	+16.21
107.23 (+1 SD)	minus	97.56 (median)	equals	+9.67
97.56 (median)	minus	97.56 (median)	equals	—
87.86 (−1 SD)	minus	97.56 (median)	equals	−9.70
73.11 (−2 SD)	minus	97.56 (median)	equals	−24.45

Latest				
Child age 4				
105.52 (+2 SD)	minus	85.80 (median)	equals	+19.72
99.57 (+1 SD)	minus	85.80 (median)	equals	+13.77
85.80 (median)	minus	85.80 (median)	equals	—
72.62 (−1 SD)	minus	85.80 (median)	equals	−13.18
59.08 (−2 SD)	minus	85.80 (median)	equals	−26.72

Step IV

To estimate the proportion of variance family environment accounts for, there must be an estimate of the cognitive value of the home at various performance levels. I assume:

(1) that before the matching of an individual's genes with environment begins, family determines virtually all of performance variance. That cannot be literally true in that genes must have some direct effect on brain physiology and contribute something to performance even early on. But I think trends will show that it is almost true – wait for the data.

(2) that those who perform at 2 SD above the median do not contain those whose families offer a cognitive value in the bottom 30 percent (those 2 SD below would not contain those in the top 30 percent). Those at one SD above the median would not contain families in the bottom 15 percent (etc.). Naturally these estimates can be challenged but the point is not to get truly accurate proportions of family variance. It is to get a series of rough estimates that show TRENDS – what the percentage is likely to be in early childhood and at what age it is likely to fade out entirely. I think that these trends will show the value even of rough estimates.

That done we can calculate the following:

(1) If family contributed 100 percent of variance before matching and if those at various levels were distributed randomly throughout the curve, then those at 2 SD above the median should be 30 points short of adults at the target age (and this would diminish gradually as family influence fades out with age).

(2) However, eliminating 30 percent from the bottom of a normal curve lifts the SD of the remainder by 0.4967 SD. That multiplied by 15 equals 7.45 points. When this is deducted from 30 to give 22.55, we have the proper divisor to calculate how much variance is being explained: say 5 points are missing at 2 SD above the median; 5 (represents the failure to regress to the mean) divided by 22.55 equals a correlation of 0.222. And that squared (0.222 × 0.222) gives 4.93 percent of the variance explained.

(3) Eliminating 15 percent from the bottom of the curve lifts the SD of the remainder by 0.2743 SD or 4.11 points and, deducted from 15, gives *10.89* as the proper divisor at 1 SD above the median level.

The above values allow us to state exactly what the gap is between a performance level and the average cognitive quality of the home at that level:

(1) Those at a 97.73 percentile performance level enjoy family cognitive quality at only the 69th percentile. When the bottom 30 percent are eliminated, the cognitive quality of the homes has been lifted to 0.4967 SD above the median: 0.4967 above the median is essentially the 69th percentile.

(2) Those at 84th percentile performance level enjoy family cognitive quality at only about the 61st percentile. When the bottom 15 percent are eliminated, the cognitive quality of the homes has been lifted to 0.2743 SD above the median: 0.2743 above the median is essentially the 61st percentile.

Now that we have the proper divisors, let us put them to work.

Table AI3 Decline of common environment effects with age averaging all years where two or more are available

		Divisor	Correlation	Ages	Ave. cor. by age	% var.	Years available
+2 SD	**+16.22**	22.55	**0.719**	4	0.896	80.24	1992 and 2004.5
+2 SD	**+14.03**	22.55	**0.622**	7	0.836	69.89	1950.5–1975–1992–2004.5
+2 SD	**+8.82**	22.55	**0.391**	9.5	0.616	37.93	"
+2 SD	**+4.84**	22.55	**0.215**	12	0.434	18.84	"
+2 SD	**+2.31**	22.55	**0.102**	14.5	0.274	7.51	"
+2 SD	**−0.80**	22.55	**−0.035**	17	0.230	5.29	"
+2 SD	**−0.64**	22.55	**−0.028**	18–19	0.158	2.49	"
+2 SD	**−0.87**	22.55	**−0.039**	20–24	0.124	1.54	"
+1 SD	**+9.67**	10.89	**0.888**	4			1992 and 2004.5
+1 SD	**+10.85**	10.89	**0.996**	7			1950.5–1975–1992–2004.5
+1 SD	**+8.61**	10.89	**0.791**	9.5			"
+1 SD	**+6.17**	10.89	**0.567**	12			"
+1 SD	**+5.30**	10.89	**0.487**	14.5			"
+1 SD	**+3.62**	10.89	**0.332**	17			"
+1 SD	**+2.84**	10.89	**0.261**	18–19			"
+1 SD	**+2.09**	10.89	**0.192**	20–24			"
−1 SD	**−9.70**	10.89	**0.891**	4			1992 and 2004.5
−1 SD	**−8.78**	10.89	**0.806**	7			1950.5–1975–1992–2004.5
−1 SD	**−6.99**	10.89	**0.642**	9.5			"
−1 SD	**−4.85**	10.89	**0.445**	12			"
−1 SD	**−1.74**	10.89	**0.160**	14.5			"
−1 SD	**−2.89**	10.89	**0.265**	17			"

		Divisor	Correlation	Ages	Ave. cor. by age	% var.	Years available
−1 SD	**−1.72**	10.89	**0.158**	18–19			"
−1 SD	**−1.45**	10.89	**0.133**	20–24			"
−2 SD	**−24.45**	22.55	**1.084**	4			1992 and 2004.5
−2 SD	**−20.75**	22.55	**0.920**	7			1950.5– 1975–1992– 2004.5
−2 SD	**−14.41**	22.55	**0.639**	9.5			"
−2 SD	**−11.47**	22.55	**0.509**	12			"
−2 SD	**−7.85**	22.55	**0.348**	14.5			"
−2 SD	**−8.07**	22.55	**0.358**	17			"
−2 SD	**−5.41**	22.55	**0.240**	18–19			"
−2 SD	**−4.75**	22.55	**0.211**	20–24			"

With the exception of age 4, the above supplies all the values under **Wechsler Vocabulary** in Table 7b in the text.

Table AI4 Decline of common environment effects with age circa 2004.5/2007

	Points	Divisor	Correlation	Ages	Ave. cor. by age	% var.
+2 SD	**+19.72**	22.55	**0.875**	4	1.134	128.48
+2 SD	**+8.91**	22.55	**0.395**	7	0.688	47.30
+2 SD	**+4.45**	22.55	**0.197**	9.5	0.463	21.45
+2 SD	**+1.91**	22.55	**0.085**	12	0.356	12.67
+2 SD	**+0.75**	22.55	**0.033**	14.5	0.197	3.89
+2 SD	**+0.25**	22.55	**0.011**	17	0.278	7.71
+2 SD	**−2.25**	22.55	**−0.100**	18–19	0.234	5.47

(continued)

Table AI4 Decline of common environment effects with age circa 2004.5/2007 (*continued*)

	Points	Divisor	Correlation	Ages	Ave. cor. by age	% var.
+2 SD	−1.25	22.55	−0.056	20–24	0.213	4.54
+2 SD	−2.00	22.55	−0.089	25–29	0.135	1.82
+2 SD	−1.75	22.55	−0.078	30–34	0.067	0.45
+2 SD	−1.50	22.55	−0.067	35–44	0.023	0.05
+1 SD	+13.77	10.89	1.264	4		
+1 SD	+9.52	10.89	0.874	7		
+1 SD	+7.48	10.89	0.687	9.5		
+1 SD	+5.77	10.89	0.530	12		
+1 SD	+4.42	10.89	0.406	14.5		
+1 SD	+5.25	10.89	0.482	17		
+1 SD	+4.75	10.89	0.436	18–19		
+1 SD	+3.75	10.89	0.344	20–24		
+1 SD	+2.00	10.89	0.184	25–29		
+1 SD	+0.25	10.89	0.023	30–34		
+1 SD	−0.50	10.89	−0.046	35–44		
−1 SD	−13.18	10.89	1.210	4		
−1 SD	−8.77	10.89	0.805	7		
−1 SD	−6.23	10.89	0.572	9.5		
−1 SD	−5.02	10.89	0.461	12		
−1 SD	−1.68	10.89	0.154	14.5		
−1 SD	−4.18	10.89	0.384	17		
−1 SD	−4.11	10.89	0.377	18–29		
−1 SD	−3.96	10.89	0.364	20–24		
−1 SD	−3.14	10.89	0.288	25–29		
−1 SD	−2.32	10.89	0.213	30–34		
−1 SD	−1.50	10.89	0.138	35–44		
−2 SD	−26.72	22.55	1.185	4		
−2 SD	−15.25	22.55	0.676	7		
−2 SD	−8.96	22.55	0.397	9.5		
−2 SD	−7.75	22.55	0.344	12		
−2 SD	−4.42	22.55	0.196	14.5		
−2 SD	−5.25	22.55	0.233	17		

	Points	Divisor	Correlation	Ages	Ave. cor. by age	% var.
−2 SD	**−5.00**	22.55	**0.222**	18–19		
−2 SD	**−4.50**	22.55	**0.200**	20–24		
−2 SD	**−3.50**	22.55	**0.155**	25–29		
−2 SD	**−2.50**	22.55	**0.111**	30–34		
−2 SD	**−1.50**	22.55	**0.067**	35–44		

With the exception of age 4, the above supplies all the values under **Wechsler Vocabulary** in Table 8b in the text.

Appendix II: Stanford-Binet Vocabulary

Sources

Thorndike, R. L., Hagen, E. P., and Sattler, J. M. (1986). *Stanford-Binet Intelligence Scale: Fourth Edition.* Chicago: Riverside. **(SB-4; 1985)**

Roid, G. H. (2003). *Stanford-Binet Intelligence Scales: Fifth Edition.* Itasca, IL: Riverside. **(SB-5; 2001)**

Table 5 in the main text uses Stanford-Binet 5 (2001) Vocabulary to provide the crucial data, data showing that there is a "pattern of progressive gaps" between earlier ages and the target age (ages 50–59). Therefore, I will first supply the raw data and calculations on which Table 5 is based. I do not use levels 1 (-3 SD) or 19 ($+3$ SD) from the manual because they are not a true measure, in the sense that they give no average score, rather they lump together all raw scores that exhaust their theoretical range. Therefore, they give no indication of either the gaps at those levels or of the variance of raw scores over the whole curve. To get the latter, subtract the raw score at level 2 from that at level 18 (for example, at the target age, $53 - 26 = 27$).

Table AII1 SB 2001 Vocabulary: progressive rise of score gaps by age (between earlier ages and the target age – data and calculations) T = target age (50–59)

	T	17–19 = **18**	gap	16.16 + 15.83 = **16**		gap	14.16 + 13.83 = **14**		gap
2	26	25.5	0.5	24.5	24.5 = 24.5	1.5	23.5	22.5 = 23	3.0
3	28.5	27.5	1.0	26.5	26.5 = 26.5	2.0	25.5	24.5 = 25	3.5
4 (−2 SD)	30.5	29.5	1.0	28.5	28.5 = 28.5	2.0	27.5	26.5 = 27	3.5
5	33	31.5	1.5	30.5	30.5 = 30.5	2.5	29.5	28.5 = 29	4.0
6	35.5	33.5	2.0	32.5	32.5 = 32.5	3.0	31.5	30 = 30.75	4.75
7 (−1 SD)	37.5	35.5	2.0	34.5	34.5 = 34.5	3.0	33.5	31.5 = 32.5	5.0
8	40	37.5	2.5	36.5	36.5 = 36.5	3.5	35	33.5 = 34.25	5.75
9	42.5	39.5	**3.0**	38.5	38.5 = 38.5	**4.0**	36.5	35.5 = 36	**6.5**
10 (med.)	44.5	41.5	**3.0**	41	40.5 = 40.75	**3.75**	38.5	37.5 = 38	**6.5**
11	46.5	43.5	**3.0**	43.5	42.5 = 43.0	**3.5**	40.5	39.5 = 40	**6.5**
12	49	46	3.0	45.5	44 = 44.75	4.25	42.5	41.5 = 42	7.0
13 (+1 SD)	51.5	48.5	3.0	47.5	45.5 = 46.5	5.0	44.5	43.5 = 44	7.5
14	53.5	50.5	3.0	49.5	47.5 = 48.5	5.0	46.5	45.5 = 46	7.5
15	56	52.5	3.5	51.5	49.5 = 50.5	5.5	48.5	47.5 = 48	8.0
16 (+2 SD)	58.5	54.5	4.0	53.5	51.5 = 52.5	6.0	50.5	49 = 49.75	8.75
17	60.5	56.5	4.0	55.5	53.5 = 54.5	6.0	52.5	50.5 = 51.5	9.0
18	63	58.5	4.5	57.5	56 = 56.75	6.25	54.5	53 = 53.75	9.25
T/B	**37**	**33**	**(4.0)**	33	31.5	**32.25 (4.75)**	31	30.5	**30.75 (6.25)**

(*continued*)

Table AII1 SB 2001 Vocabulary: progressive rise of score gaps by age (between earlier ages and the target age – data and calculations) T = target age (50–59) (*continued*)

	T	12.16 + 11.83 = **12**		gap	10.16 + 9.83 = **10**		gap
2	26	21.5	21.5 = 21.5	4.5	20.5	19.5 = 20	6.0
3	28.5	23.5	23.0 = 23.25	5.25	22.5	21 = 21.75	6.75
4 (−2 SD)	30.5	25.5	24.5 = 25.0	5.5	24.5	22.5 = 23.5	7.0
5	33	27.5	26.5 = 27.0	6.5	26	24.5 = 25.25	7.75
6	35.5	29.5	28.5 = 29.0	6.5	27.5	26.5 = 27	8.5
7 (−1 SD)	37.5	31	30.5 = 30.75	6.75	29.5	28 = 28.75	8.75
8	40	32.5	32.0 = 32.25	7.75	31	29.5 = 30.25	9.75
9	42.5	34.5	33.5 = 34	**8.5**	32.5	31.5 = 32	**10.5**
10 (med.)	44.5	36.5	35.5 = 36	**8.5**	34.5	33.5 = 34	**10.5**
11	46.5	38.5	37.5 = 38	**8.5**	36.5	35 = 35.75	**10.75**
12	49	40.5	39.5 = 40	9.0	38	36.5 = 37.25	11.75
13 (+1 SD)	51.5	42.5	41.5 = 42	9.5	39.5	38.5 = 39	12.5
14	53.5	44	43 = 43.5	10.0	41.5	40.5 = 41	12.5
15	56	45.5	44.5 = 45	11.0	43	42 = 42.5	13.5
16 (+2 SD)	58.5	47.5	46.5 = 47	11.5	44.5	43.5 = 44	14.5
17	60.5	49.5	48.5 = 49	11.5	46.5	45.5 = 46	14.5
18	63	51.5	50.5 = 51	12.0	48.5	47.5 = 48	15.0
T/B	**37**	30	29	**29.5 (7.5)**	28	28	**28 (9.0)**

188

	T	8.16 + 7.83 = 8		gap	6.16 + 5.83 = 6		gap
2	26	17	16 = 16.5	9.5	11	10.5 = 10.75	15.25
3	28.5	18.5	17.5 = 18.0	10.5	12.5	12.5 = 12.5	16.00
4 (−2 SD)	30.5	20.5	19 = 19.75	10.75	14.5	14 = 14.25	16.25
5	33	22	20.5 = 21.25	11.75	16	15.5 = 15.75	17.25
6	35.5	23.5	22.5 = 23.0	12.5	17.5	17 = 17.25	18.25
7 (−1 SD)	37.5	25.5	24 = 24.75	12.75	19	18.5 = 18.75	18.75
8	40	27	25.5 = 26.25	13.75	20.5	20 = 20.25	19.75
9	42.5	28.5	27 = 27.75	**14·75**	22.5	21.5 = 22.0	**20.50**
10 (med.)	44.5	30.5	28.5 = 29.5	**15.0**	24	23 = 23.5	**21.00**
11	46.5	32	30.5 = 31.25	**15·25**	25.5	24.5 = 25.0	**21.50**
12	49	33.5	32 = 32.75	16.25	27	26 = 26.5	22.50
13 (+1 SD)	51.5	35.5	33.5 = 34.5	17.00	28.5	27.5 = 28.0	23.50
14	53.5	37	35.5 = 36.25	17.25	30	29.5 = 29.75	23.75
15	56	38.5	37 = 37.75	18.25	31.5	31 = 31.25	24.25
16	58.5	40	38.5 = 39.75	18.75	33.5	32.5 = 33.0	25.50
17	60.5	41.5	40.5 = 41	19.5	35	34 = 34.5	26.00
18 (+2 SD)	63	43	42 = 42.5	20.5	37	36 = 36.5	26.50
T/B	37	26	26	**26 (11.0)**	26	25.5	**25.75 (11.25)**

(continued)

Table AII1 SB 2001 Vocabulary: progressive rise of score gaps by age (between earlier ages and the target age – data and calculations) T = target age (50-59) (*continued*)

	T	4.08 + 3.92 = 4		gap	2.08 = 2	gap
2	26	6	5.5 = 5.75	20.25	1	25
3	28.5	7.5	7 = 7.25	21.25	2	26.5
4 (−2 SD)	30.5	9	8.5 = 8.75	21.75	3	27.5
5	33	10.5	10 = 10.25	22.75	4	29
6	35.5	12	11 = 11.5	24.00	5	30.5
7 (−1 SD)	37.5	13	12.5 = 12.75	24.75	6	31.5
8	40	14.5	14 = 14.25	25.75	6.5	33.5
9	42.5	16	15.5 = 15.75	**26.75**	7	**35.5**
10 (med.)	44.5	17.5	17 = 17.25	**27.25**	8	**36.5**
11	46.5	19	18.5 = 18.75	**27.75**	9	**37.5**
12	49	20	20 = 20	29	10	39
13 (+1 SD)	51.5	21.5	21 = 21.25	30.25	11	40.5
14	53.5	23	22.5 = 22.75	30.75	12.5	41
15	56	24.5	24 = 24.25	31.75	14	42
16 (+2 SD)	58.5	26	25.5 = 25.75	32.75	15	43.5
17	60.5	27	27 = 27	33.5	16	44.5
18	63	29	28.5 = 28.75	34.25	17.5	45.5
T/B	37	23	23	23 (**14.00**)		**16.5 (20.5)**

Now I can pass on to give the usual data on which the estimates of the disadvantage/advantages at each performance level is based.

Step I

SB Vocabulary (or Verbal Knowledge) data: raw scores by age at various performance levels. Values in bold emphasized as key in the computations that follow.

					SB-4 Vocabulary					

$2 = 2/0-2/3.5$; $2.5 = 2/3.5-2/7.5$; $3 = 2/11.5-3/3.5$; $4 = 3/11.5-4/3.5$; $6.75 = 6/5.5-6/11.5$; $9.25 = 8/11.5-9/5.5$; $11.5 = 10/11.5-11/11.5$; $14.5 = 13/11.5-14/11.5$; $17.5 = 16/11.5-17/11.5$

	2	2.5	3	4	6.75	9.25	11.5	14.5	17.5	**18–24**
+2 SD	12.5	13.67	16.33	19	23.75	29.5	34.5	37.5	41.67	44
+1 SD	8.5	10.33	13	16	21.33	26	30	33.5	38	40.5
Med.	4.5	6	9	13	18.5	22.5	26	28.5	33.5	36
−1 SD	1	2	4.5	9	15	19	22	24.5	28	29.5
−2 SD	—	—	0	3.5	10	15	18	20.5	23.5	25
−3 SD				(−2)	3		11			18

					SB-5 Verbal Knowledge					

$2 = 2/0-2/1$; $2.5 = 2/4-2/5$; $3 = 3/0-3/1$; $4 = 4/0-4/1$; $6.75 = 6/8-6/11$; $9.25 = 9/0-9/3$; $11.5 = 11/4-11/7$; $14.5 = 14/4-14/7$; $18 = 17/0-18/11$

	2	2.5	3	4	6.75	9.25	11.5	14.5	18	20–24
+3 SD			(26)	(31)	(41.5)	(47)	(52)	(56.5)	(60.5)	
+2 SD	15	17.5	21	26	36.5	42	46.5	50.5	54.5	55.5
+1 SD	11	14	17	21.5	31.5	37	41.5	44.5	48.5	49
Med.	8	10	13	17.5	27	32	35.5	38.5	41.5	42.5
−1 SD	6	6.5	9	13	22.5	27	30.5	33.5	35.5	36.5
−2 SD	3	3	5	9	17.5	22	24.5	27.5	29.5	29.5
−3 SD			(1)	(5)	(12)	(17)	(18.5)		(23.5)	(22.5)

	25-29	30-34	35-39	40-45	45-49	**50-54**	**55-59**	60-64	65-69
+2 SD	56.5	57.5	57.5	57.5	57.5	58.5	58.5	58.5	58.5
+1 SD	49.5	50.5	50.5	51	51	51.5	51.5	51.5	51.5
Med.	43.5	43.5	43.5	44.5	44.5	44.5	44.5	44.5	44
−1 SD	36.5	37.5	37.5	37.5	37.5	37.5	37.5	37	36.5
−2 SD	30.5	30.5	30.5	30.5	30.5	30.5	30.5	30	29.5
−3 SD						(23.5)			

To adjust the SB-4 result, to compensate for the fact that earlier ages had to be normed on ages 18–24, I need to calculate what difference it would make had the SB-5 also been normed on ages 18–24. Therefore, I got raw score values by averaging the ones for both ages 18 and 20–24 from the SB-5 data.

+2 SD 55   +1 SD 48.75   Med.   42   −1 SD 36   −2 SD 29.5   −3 SD (23)

Step II

How to get the values by level (+2 to −2 SD).

An example should make this clear – take the value for age 11.5 in the first row:

	SB-4 data	
	11.5	18–24
+2 SD	**34.5**	44
+1 SD	30	40.5
Med.	26	**36**
−1 SD	22	**29.5**
−2 SD	18	25
−3 SD		18

36 − 34.5 = 1.5; 1.5 ÷ 6.5 (the distance between 36 and 29.5) = 0.23 SD; adding 2 SD to that gives 2.23 SD; 2.23 × 15 = **33.46** IQ difference.

(1) Since the SB-4 data has age 18–24 as its highest age, I have adjusted for the fact that IQ actually peaks at a later age. This was done in three steps:

   (i) We see what effect it would have on the SB-5 results if they were normed on that age (rather than ages 50–59).

   (ii) The calculation (done by the same method) showed that this would have lowered the value for age 11.5 by 5.71 points (from the table: 25.71 − 20.00 = 5.71). Therefore that amount was added on to the SB-4 result: 33.46 + 5.71 = **39.17**, which appears in the table as "(1885) adjusted."

(2) When you have age 11.5 normed on the target adult age (50–59), you can norm earlier years on it, and it will link them to the target adult years. For example, age 6.75 was normed on age 11.5 using the same method:

	SB-4 data	
	6.75	11.5
+2 SD	**23.75**	34.5
+1 SD	21.33	30
Med	18.5	**26**
−1 SD	15	**22**
−2 SD	10	18
−3 SD		11

26 − 23.75 = 2.25; 2.25 ÷ 4 (the distance between 26 and 22) = 0.5625 SD; adding 2 SD to that gives 2.5625 SD; 2.5625 × 15 = **38.44** IQ difference.

(3) 39.17 + 38.44 gives **77.61** as 1985 adjusted, age 6.75 normed on ages 50–59.

All values below (Table AII2A) were calculated in this way.

Once you have a value comparing age 6.75 and the adult maximum, you can use that age as a link by norming earlier ages on age 6.75. For example, in the first row in Table AII2B, add 40.60 to 77.61 and you get 118.21 for age 3 compared to ages 50–59.

Table AII2A Adult versus child IQ differences at two times at five IQ levels (all normed on adult curve)

Child age	Compared to adults					To age 11.5		To adults		Adult
	11.5	14.5	17.5/18	20–24	25–29	6.75	9.25	6.75	9.25	
+2 SD (1985)	33.46	25.00	10.00	—	—	38.44	16.88	71.90	50.34	18–24
(1985) adjusted	**39.17**	**32.14**	**18.57**	—	—	**38.44**	**16.88**	**77.61**	**56.05**	**50–59**
+2SD (2001)	**25.71**	**17.14**	**8.57**	**6.43**	**4.29**	**27.50**	**13.50**	**53.21**	**39.21**	**50–59**
(2001) adjusted	20.00	10.80	0.00	—	—	—	—	—	—	18–24
Average of bold	**32.44**	**24.64**	**13.57**	—	—	—	—	**65.41**	**47.63**	
+1 SD (1985)	28.85	20.77	8.33	—	—	32.51	15.00	61.36	43.85	18–24
(1985) adjusted	**34.03**	**26.23**	**14.76**	—	—	**32.51**	**15.00**	**66.54**	**49.03**	**50–59**
+1 SD (2001)	**21.43**	**15.00**	**6.43**	**5.36**	**4.29**	**27.00**	**11.25**	**48.43**	**32.68**	**50–59**
(2001) adjusted	16.25	9.44	0.00	—	—	—	—	—	—	18–24
Average of bold	**27.73**	**20.62**	**10.60**	—	—	—	—	**57.49**	**40.86**	
Median (1985)	26.67	18.33	5.77	—	—	28.13	13.13	54.80	39.80	18–24
(1985) adjusted	**29.96**	**22.44**	**12.20**	—	—	**28.13**	**13.13**	**58.09**	**43.09**	**50–59**
Median (2001)	**19.29**	**12.86**	**6.43**	**4.29**	**2.14**	**23.75**	**10.50**	**43.04**	**29.79**	**50–59**
(2001) adjusted	16.00	8.75	0.00	—	—	—	—	—	—	18–24
Average of bold	**24.63**	**17.65**	**9.32**	—	—	—	—	**50.57**	**36.44**	
−1 SD (1985)	21.43	16.07	5.00	—	—	21.43	11.25	42.86	32.68	18–24
(1985) adjusted	**23.74**	**18.87**	**9.29**	—	—	**21.43**	**11.25**	**45.17**	**34.99**	**50–59**
−1 SD (2001)	**15.00**	**8.57**	**4.29**	**2.14**	**2.14**	**20.00**	**8.75**	**35.00**	**23.85**	**50–59**
(2001) adjusted	12.69	5.77	0.00	—	—	—	—	—	—	18–24
Average of bold	**19.37**	**13.72**	**6.79**	—	—	—	—	**40.09**	**29.42**	
−2 SD (1985)	15.00	9.64	3.21	—	—	17.14	6.43	32.14	21.43	18–24
(1985) adjusted	**16.32**	**11.45**	**5.35**	—	—	**17.14**	**6.43**	**33.46**	**22.75**	**50–59**
−2 SD (2001)	**12.86**	**6.43**	**2.14**	**2.14**	**2.14**	**17.50**	**6.25**	**30.36**	**19.11**	**50–59**
(2001) adjusted	11.54	4.62	0.00	—	—	—	—	—	—	18–24
Average of bold	**14.59**	**8.94**	**3.75**	—	—	—	—	**31.90**	**20.93**	

Table AII2B Adult versus (young) child IQ differences at two times at five IQ levels (all normed on adult curve)

Compared to:						Adults			Adult age
Normative age	4	4	4	6.75	6.75	6.75	3	4	
Child age	2	2.5	3	4	3				
+2 SD 1985	31.88	26.65	13.35	27.35	**40.60**	**77.61**	**118.21**	**104.96**	50–59
+2 SD 2001	38.33	30.00	16.88	33.33	50.21	53.21	103.42	86.54	50–59
Average	**35.11**	**28.33**				**65.41**	**110.82**	**95.75**	
+1 SD 1985	31.36	25.00	15.00	25.71	40.71	66.54	107.25	92.95	50–59
+1 SD 2001	37.50	26.67	16.67	33.00	49.67	48.43	98.10	81.43	50–59
Average	**34.43**	**25.84**				**57.49**	**102.68**	**87.19**	
Med. 1985	27.27	23.18	15.00	21.00	36.00	58.09	94.09	79.09	50–59
Med. 2001	33.75	26.25	15.00	30.00	45.00	43.04	88.04	73.04	50–59
Average	**30.51**	**24.72**				**50.57**	**91.07**	**76.07**	
−1 SD 1985	21.82	19.09	12.27	17.14	29.41	45.17	74.58	62.31	50–59
−1 SD 2001	26.25	24.38	15.00	27.27	42.27	35.00	77.27	62.27	50–59
Average	**24.04**	**21.74**				**40.09**	**75.93**	**62.29**	
−2 SD 19.85	—	—	9.55	13.93	23.48	33.46	56.94	47.39	50–59
−2 SD 2001	—	—	15.00	23.18	38.18	30.34	68.52	53.52	50–59
Average						**31.90**	**62.73**	**50.46**	

Step III

Subtract the difference at the median from the difference at all other levels.

Table AII3 Adult versus youth Vocabulary gaps: how much do gaps at levels above/below median differ from those at median? 1985 and 2001 averaged

		Child age 3		
110.82 (+2 SD)	minus	91.07 (median)	equals	+19.75
102.68 (+1 SD)	minus	91.07 (median)	equals	+11.61
91.07 (median)	minus	91.07 (median)	equals	—
75.93 (−1 SD)	minus	91.07 (median)	equals	−15.14
62.73 (−2 SD)	minus	91.07 (median)	equals	−28.34
		Child age 4		
95.75 (+2 SD)	minus	76.07 (median)	equals	+19.68
87.39 (+1 SD)	minus	76.07 (median)	equals	+11.32
76.07 (median)	minus	76.07 (median)	equals	—
62.29 (−1 SD)	minus	76.07 (median)	equals	−13.78
50.46 (−2 SD)	minus	76.07 (median)	equals	−25.61
		Child age 6.75		
65.41 (+2 SD)	minus	50.57 (median)	equals	+14.84
57.49 (+1 SD)	minus	50.57 (median)	equals	+6.92
50.57 (median)	minus	50.57 (median)	equals	—
40.09 (−1 SD)	minus	50.57 (median)	equals	−10.48
31.90 (−2 SD)	minus	50.57 (median)	equals	−18.67
		Child age 9.25		
47.63 (+2 SD)	minus	36.44 (median)	equals	+11.19
40.86 (+1 SD)	minus	36.44 (median)	equals	+4.42
36.44 (median)	minus	36.44 (median)	equals	—
29.42 (−1 SD)	minus	36.44 (median)	equals	−7.02
20.93 (−2 SD)	minus	36.44 (median)	equals	−15.51

Child age 11.5

32.44 (+2 SD)	minus	24.63 (median)	equals	+7.81
27.73 (+1 SD)	minus	24.63 (median)	equals	+3.14
24.63 (median)	minus	24.63 (median)	equals	—
13.72 (−1 SD)	minus	24.63 (median)	equals	−5.26
14.59 (−2 SD)	minus	24.63 (median)	equals	−10.04

Child age 14.5

24.64 (+2 SD)	minus	17.65 (median)	equals	+6.99
20.62 (+1 SD)	minus	17.65 (median)	equals	+2.97
17.65 (median)	minus	17.65 (median)	equals	—
13.72 (−1 SD)	minus	17.65 (median)	equals	−3.93
8.94 (−2 SD)	minus	17.65 (median)	equals	−8.71

Child age 17.5

13.57 (+2 SD)	minus	9.32 (median)	equals	+4.25
10.62 (+1 SD)	minus	9.32 (median)	equals	+1.30
9.32 (median)	minus	9.32 (median)	equals	—
6.79 (−1 SD)	minus	9.32 (median)	equals	−2.53
3.75 (−2 SD)	minus	9.32 (median)	equals	−5.57

Latest (2001)

Child age 3

103.42 (+2 SD)	minus	88.04 (median)	equals	+15.38
98.10 (+1 SD)	minus	88.04 (median)	equals	+10.06
88.04 (median)	minus	88.04 (median)	equals	—
77.27 (−1 SD)	minus	88.04 (median)	equals	−10.77
68.52 (−2 SD)	minus	88.04 (median)	equals	−9.52

Child age 4

86.54 (+2 SD)	minus	73.04 (median)	equals	+13.50
81.43 (+1 SD)	minus	73.04 (median)	equals	+8.39
73.04 (median)	minus	73.04 (median)	equals	—
62.27 (−1 SD)	minus	73.04 (median)	equals	−10.77
53.52 (−2 SD)	minus	73.04 (median)	equals	−19.52

(continued)

Table AII3 Adult versus youth Vocabulary gaps: how much do gaps at levels above/below median differ from those at median? 1985 and 2001 averaged (*continued*)

		Child age 6.75		
53.21 (+2 SD)	minus	43.04 (median)	equals	+10.17
48.43 (+1 SD)	minus	43.04 (median)	equals	+5.39
43.04 (median)	minus	43.04 (median)	equals	—
35.00 (−1 SD)	minus	43.04 (median)	equals	−8.04
30.34 (−2 SD)	minus	43.04 (median)	equals	−12.70
		Child age 9.25		
39.21 (+2 SD)	minus	29.29 (median)	equals	+9.92
32.68 (+1 SD)	minus	29.29 (median)	equals	+3.39
29.79 (median)	minus	29.29 (median)	equals	—
23.85 (−1 SD)	minus	29.29 (median)	equals	−5.44
19.11 (−2 SD)	minus	29.29 (median)	equals	−10.18
		Child ages 11.5		
25.71 (+2 SD)	minus	19.29 (median)	equals	+6.42
21.43 (+1 SD)	minus	19.29 (median)	equals	+2.14
19.29 (median)	minus	19.29 (median)	equals	—
15.00 (−1 SD)	minus	19.29 (median)	equals	−4.29
12.86 (−2 SD)	minus	19.29 (median)	equals	−6.43
		Child ages 14.5		
17.14 (+2 SD)	minus	12.86 (median)	equals	+4.28
15.00 (+1 SD)	minus	12.86 (median)	equals	+2.14
12.86 (median)	minus	12.86 (median)	equals	—
8.57 (−1 SD)	minus	12.86 (median)	equals	−4.29
6.43 (−2 SD)	minus	12.86 (median)	equals	−6.43
		Age 18		
8.57 (+2 SD)	minus	6.43 (median)	equals	+2.14
6.43 (+1 SD)	minus	6.43 (median)	equals	0.00
6.43 (median)	minus	6.43 (median)	equals	—
4.29 (−1 SD)	minus	6.43 (median)	equals	−2.14
2.14 (−2 SD)	minus	6.43 (median)	equals	−4.29

Ages 20–24				
6.43 (+2 SD)	minus	4.29 (median)	equals	+2.14
5.36 (+1 SD)	minus	4.29 (median)	equals	+1.07
4.29 (median)	minus	4.29 (median)	equals	—
2.14 (−1 SD)	minus	4.29 (median)	equals	−2.15
2.14 (−2 SD)	minus	4.29 (median)	equals	−2.15
Ages 25–29				
4.29 (+2 SD)	minus	2.14 (median)	equals	+2.15
4.29 (+1 SD)	minus	2.14 (median)	equals	+2.15
2.14 (median)	minus	2.14 (median)	equals	—
2.14 (−1 SD)	minus	2.14 (median)	equals	0.00
2.14 (−2 SD)	minus	2.14 (median)	equals	0.00

Step IV

To estimate the proportion of variance family environment accounts for, there must be an estimate of the cognitive value of the home at various performance levels. See Appendix I as a guide for the derivation of the devisors used below.

Table AII4 Decline of common environment effects with age, 1985 and 2001 averaged

	Points	Divisor	Correlation	Ages	Ave. Cor. by age	% var.
+2 SD	**+19.75**	22.55	**0.876**	3	1.147	131.59
+2 SD	**+19.68**	22.55	**0.873**	4	1.064	113.21
+2 SD	**+14.84**	22.55	**0.658**	6.75	0.771	59.44
+2 SD	**+11.19**	22.55	**0.496**	9.25	0.559	31.23
+2 SD	**+7.81**	22.55	**0.346**	11.5	0.391	15.27
+2 SD	**+6.99**	22.55	**0.310**	14.5	0.333	11.07
+2 SD	**+4.25**	22.55	**0.188**	17.5/18	0.197	3.87
+1 SD	**+11.61**	10.89	**1.066**	3		
+1 SD	**+11.32**	10.89	**1.039**	4		

(*continued*)

Table AII4 Decline of common environment effects with age, 1985 and 2001 averaged (*continued*)

	Points	Divisor	Correlation	Ages	Ave. Cor. by age	% var.
+1 SD	**+6.92**	10.89	**0.635**	6.75		
+1 SD	**+4.42**	10.89	**0.406**	9.25		
+1 SD	**+3.14**	10.89	**0.288**	11.5		
+1 SD	**+2.97**	10.89	**0.273**	14.5		
+1 SD	**+1.30**	10.89	**0.119**	17.5/18		
−1 SD	−15.14	10.89	**1.390**	3		
−1 SD	−13.17	10.89	**1.209**	4		
−1 SD	−10.48	10.89	**0.963**	6.75		
−1 SD	−7.02	10.89	**0.645**	9.25		
−1 SD	−5.26	10.89	**0.483**	11.5		
−1 SD	−3.93	10.89	**0.361**	14.5		
−1 SD	−2.53	10.89	**0.232**	17.5/18		
−2 SD	−28.34	22.55	**1.257**	3		
−2 SD	−25.61	22.55	**1.136**	4		
−2 SD	−18.67	22.55	**0.828**	6.75		
−2 SD	−15.51	22.55	**0.688**	9.25		
−2 SD	−10.04	22.55	**0.445**	11.5		
−2 SD	−8.71	22.55	**0.386**	14.5		
−2 SD	−5.57	22.55	**0.247**	17.5/18		

With the exception of ages 3 and 4, Table AII4 supplies all the values under **Stanford-Binet Vocabulary** in Table 7b in the text.

Table AII5 supplies all the values in the text for Figure 1, and, with the exception of ages 3 and 4, all those under **Stanford-Binet Vocabulary** in Table 8b.

Table AII5 Decline of common environment effects with age 2001

	Points	Divisor	Correlation	Ages	Ave. cor. by age	% var. family	% var. uncommon	% var. genes	Dutch genes
+2 SD	+15.38	22.55	0.682	3	0.865	74.88	12.56	12.56	—
+2 SD	+13.50	22.55	0.599	4	0.806	64.96	18.00	17.04	22 (5)
+2 SD	+10.17	22.55	0.451	6.75	0.562	31.57	18.00	50.43	40 (7)
+2 SD	+9.92	22.55	0.440	9.25	0.426	18.13	18.00	63.87	54 (10)
+2 SD	+6.42	22.55	0.285	11.5	0.290	8.42	18.00	73.58	85 (12)
+2 SD	+4.28	22.55	0.190	14.5	0.267	7.12	18.00	74.88	—
+2 SD	+2.14	22.55	0.095	18	0.121	1.46	18.00	80.54	82 (18)
+2 SD	+2.14	22.55	0.095	20–24	0.121	1.46	18.00	80.54	—
+2 SD	+2.15	22.55	0.095	25–29	0.073	0.53	18.00	81.47	88 (26)
+1 SD	+10.06	10.89	0.924	3					
+1 SD	+8.39	10.89	0.770	4					
+1 SD	+5.39	10.89	0.495	6.75					
+1 SD	+3.39	10.89	0.311	9.25					
+1 SD	+2.14	10.89	0.197	11.5					
+1 SD	+2.14	10.89	0.197	14.5					
+1 SD	0.00	10.89	0.000	17.5/18					
+1 SD	+1.07	10.89	0.098	20–24					

(*continued*)

Table AII5 Decline of common environment effects with age 2001 (*continued*)

	Points	Divisor	Correlation	Ages	Ave. cor. by age	% var. family	% var. uncommon	% var. genes	Dutch genes
+1 SD	+2.15	10.89	0.197	25–29					
–1 SD	–10.77	10.89	0.989	3					
–1 SD	–10.77	10.89	0.989	4					
–1 SD	–8.04	10.89	0.738	6.75					
–1 SD	–5.44	10.89	0.500	9.25					
–1 SD	–4.29	10.89	0.394	11.5					
–1 SD	–4.29	10.89	0.394	14.5					
–1 SD	–2.14	10.89	0.197	17.5/18					
–1 SD	–2.15	10.89	0.197	20–24					
–1 SD	0.00	10.89	0.000	25–29					
–2 SD	–19.52	22.55	0.866	3					
–2 SD	–19.52	22.55	0.866	4					
–2 SD	–12.70	22.55	0.563	6.75					
–2 SD	–10.18	22.55	0.451	9.25					
–2 SD	–6.43	22.55	0.285	11.5					
–2 SD	–6.43	22.55	0.285	14.5					
–2 SD	–0.29	22.55	0.190	17.5/18					
–2 SD	–2.15	22.55	0.095	20–24					
–2 SD	0.00	22.55	0.000	25–29					

Appendix III: Raven's Progressive Matrices

Sources

Raven, J. C. (1941). Standardization of progressive matrices. *British Journal of Medical Psychology* 19: 137–50. **Contains SPM 1938.**

Foulds, G. A., and Raven, J. C. (1948). Normal changes in the mental abilities of adults as age advances. *Journal of Mental Science* 94: 133–42. **Contains SPM normed on adults 1942.** Although the data were collected over a number of years, they are always referred to as the "1942 UK adult norms."

Raven, J. C., Court, J. H., and Raven, J. (1976). *Manual for Raven's Progressive Matrices and Vocabulary Scales.* London: Lewis. **Contains CPM 1949** (sometimes dated as 1947).

Raven, J. C., Court, J. H., and Raven, J. (1986). *Manual for Raven's Progressive Matrices and Vocabulary Scales.* London: H. K. Lewis. **Contains 1979 SPM and 1982 CPM.**

Raven, J., Raven, J. C., and Court, J. H. (2003, updated 2004). *Manual for Raven's Progressive Matrices and Vocabulary Scales.* San Antonio, TX: Harcourt. Table APM XIV **links ages 15.5 with ages 20–70 for APM 1992 raw scores.**

Raven, J., Rust, J., and Squire, A. (2008a). *Manual: Coloured Progressive Matrices and Crichton Vocabulary Scales.* London: Pearson. **Contains CPM 2007.**

Raven, J., Rust, J., and Squire, A. (2008b). *Raven's Standard Progressive Matrices (SPM) and Raven's Standard Progressive Matrices Plus (SPM Plus).* London: Pearson. **Contains SPM Plus 2008.**

The most recent data (linking 2007 CPM with 2008 SPM with 1992 APM)

First, I will do the most recent data for all results throughout.

Step I. Data: raw scores by age at various performance levels.

Step II. How to get the values by SD level: given the great interest of this data, I will detail my calculations throughout.

Step III. Subtract the difference at the median from the difference at all other levels.

Standard Progressive Matrices Plus (2008)

First, I norm age 7.5 on 9.5 and then, age 9.5 on 15.5 to get a cumulative total (for 7.5 on 15.5). This version of the Progressive Matrices added new items of greater difficulty, which means that the raw scores by percentile are not analogous to previous standardizations of the SPM. But the test is similar in kind. The values in bold are relevant to the first norming (7.5/9.5), the values in brackets relevant to the second (9.5/15.5).

		Age in years								
Percentile	SD	7.5	8.5	9.5	10.5	11.5	12.5	13.5	14.5	15.5
95	+1.645	**31.0**	36.5	(37.0)	38.0	38.0	41.5	41.5	41.5	43.5
90	+1.282	**29.8**	34.2	(35.3)	36.9	36.9	39.3	39.9	39.9	41.8
75	+0.674	**26.5**	30.5	(**33.0**)	34.5	35.0	36.5	37.5	37.5	(39.0)
50	—	**22.5**	26.5	(**30.5**)	32.0	32.5	33.0	34.5	34.5	(35.5)
25	−0.674	**18.5**	22.5	(**26.5**)	28.0	28.0	30.5	31.5	31.5	(32.5)
10	−1.282	14.4	18.4	**21.4**	23.8	23.8	25.9	28.7	28.7	(28.8)
5	−1.645	11.5	15.0	**18.0**	21.5	21.5	23.5	26.0	26.0	(26.5)

+1.645 SD (age 7.5) = +23.10 IQ points (normed on age 9.5)

(1) 31.0 = +1.645 SD (age 7.5); 31.0 becomes 54.17 percentile or +0.105 SD (age 9.5).

(2) +1.645 minus +0.105 = +1.54 SD or a 23.10 IQ-point gap at that level.

+1.282 SD (age 7.5) = +20.88 IQ points (normed on age 9.5)

(1) 29.8 = +1.282 SD (age 7.5); 29.8 becomes 45.625 percentile or −0.110 SD (age 9.5).
(2) +1.282 minus −0.110 = +1.392 SD or a 20.88 IQ-point gap at that level.

+0.674 SD (age 7.5) = +20.22 IQ points (normed on age 9.5)

(1) 26.5 = +0.674 SD (age 7.5); 26.5 becomes 25 percentile or −0.674 SD (age 9.5).
(2) +0.674 minus −0.674 = +1.348 SD or a 20.22 IQ-point gap at that level.

Median (age 7.5) = +16.73 IQ points (normed on age 9.5)

(1) 22.5 = 0.00 SD (age 7.5); 22.5 becomes 13. 235 percentile or −1.1155 SD (age 9.5).
(2) 0.00 minus −1.1155 = +1.1155 SD or a 16.73 IQ-point gap at that level.

−0.674 SD (age 7.5) = +12.79 (normed on age 9.5)

(1) 18.5 = −0.674 SD (age 7.5); 18.5 becomes 6.471 percentile or −1.527 SD (age 9.5).
(2) −0.674 minus −1.527 = +0.853 SD or a 12.79 IQ-point gap at that level.

1.282 SD and **−1.645 SD**: no estimates at these levels. Look at the data and you will see that the raw scores for the earlier age are off the scale of the older age. So that various ages will be comparable, I have assumed that the estimates would be identical to the lowest we have (−0.674 SD). I call these hypothetical estimates. Since values tend to rise the further we go below the median, the true values might be a bit higher.

Now subtract the median value (allow for age) from the others.

Age 7.5 normed on age 9.5
+1.645 SD = +23.10 IQ points minus +16.73 = +6.37
+1.282 SD = +20.88 IQ points minus +16.73 = +4.15
+0.674 SD = +20.22 IQ points minus +16.73 = +3.49
Median = +16.73 IQ points minus +16.73 = NIL
−0.674 SD = +12.79 IQ points minus +16.73 = −3.94
−1.282 SD = (hypothetical estimate) = −3.94
−1.645 SD = (hypothetical estimate) = −3.94

The classic pattern of family effects disadvantageous above the median and advantageous below asserts itself!

+1.645 SD (age 9.5) = +20.60 IQ points (normed on age 15.5)

(1) 37.0 = +1.645 SD(age 9.5); 37.0 becomes 60.714 percentile or +0.272 SD(age 15.5).

(2) +1.645 minus +0.272 = +1.37 SD or a 20.60 IQ-point gap at that level.

+1.282 SD (age 9.5) = +19.86 IQ points (normed on age 15.5)

(1) 35.3 = +1.282 SD (age 9.5); 35.3 becomes 48.33 percentile or −0.042 SD (age 15.5).

(2) +1.282 minus −0.042 = +1.324 SD or a 19.86 IQ-point gap at that level.

+0.674 SD (age 9.5) = +18.34 IQ points (normed on age 15.5)

(1) 33.0 = +0.674 SD(age 9.5); 26.5 becomes 29.17 percentile or −0.5485 SD (age 15.5).

(2) +0.674 minus −0.5485 = +1.2225 SD or an 18.34 IQ-point gap at that level.

Median (age 9.5) = +14.38 IQ points (normed on age 15.5)

(1) 30.5 = 0.00 SD (age 9.5); 30.5 becomes 16.89 percentile or −0.9585 SD (age 15.5).

(2) 0.00 minus −0.9585 = +0.9585 SD or a 14.38 IQ-point gap at that level.

−0.674 SD (aged 9.5) = +14.56 points (normed on age 15.5)

(1) 26.5 = −0.674 SD (age 9.5); 26.5 becomes 5.00 percentile or −1.645 SD (age 15.5).

(2) −0.674 minus −1.645 = +0.971 SD or a 14.56 IQ-point gap at that level.

−1.282 SD and **−1.645 SD**: only hypothetical estimates at these levels – see above.

Now subtract the median value (allow for age) from the others.

Age 9.5 normed on age 15.5

+1.645 SD = +20.60 IQ points minus +14.38 = +6.22
+1.282 SD = +19.86 IQ points minus +14.38 = +5.48
+0.674 SD = +18.34 IQ points minus +14.38 = +3.96
Median = +14.38 IQ points minus +14.38 = NIL
−0.674 SD = +14.56 IQ points minus +14.38 = +0.18
−1.282 SD = (hypothetical estimate) = +0.18
−1.645 SD = (hypothetical estimate) = +0.18

The classic pattern holds above the median, but there is nil effect below the median. Nonetheless, the cumulative values are "correct," though still low below the median.

Cumulative totals for age 7.5 normed on age 15.5

+1.645 SD = +6.37 plus +6.22 = +12.59
+1.282 SD = +4.15 plus +5.48 = +9.63
+0.674 SD = +3.39 plus +3.96 = +7.35
−0.674 SD = −3.94 plus +0.18 = −3.76
−1.282 SD = (−3.76)
−1.645 SD = (−3.76)

Advanced Progressive Matrices (1992)

Note: there is only one standardization here. However, scores for 15.5-year-olds really date back to 1979. Unless IQ gains over that period (1979–92) showed very different patterns of family effects, the comparison by percentile level should be valid. I make ages 18–32 (median 25) the target ages because performance peaks then (without variation) and declines thereafter. The values in bold are relevant to norming age 15.5 on age 25.

		Age in years	
Percentile	SD	15.5	18–32 (25)
95	+1.645	**27.0**	33.0
90	+1.282	**23.0**	31.0
75	+0.674	**18.0**	**27.0**
55	+0.126	14.7	**23.0**
50	—	**14.0**	22.0
29	−0.553	10.7	**18.0**
25	−0.674	**10.0**	17.0
15	−1.036	8.0	**14.0**
10	−1.282	**7.0**	12.0
6	−1.555	6.0	**10.0**
5	−1.645	**5.5**	9.0
3	−1.881	—	**7.0**
1.5	−2.170	—	**5.5**

+1.645 SD (age 15.5) = +14.565 IQ points (normed on age 25)

(1) 27.0 = +1.645 SD (age 15.5); 27.0 becomes 75.0 percentile or +0.674 SD (age 25).

(2) +1.645 minus +0.674 = +0.971 SD or a 14.565 IQ-point gap at that level.

+1.282 SD (age 15.5) = +17.34 IQ points (normed on age 25)

(1) 23 = +1.282 SD (age 7.5); 23.0 becomes 55.0 percentile or +0.126 SD (age 25).

(2) +1.282 minus +0.126 = +1.156 SD or a 17.34 IQ-point gap at that level.

+0.674 SD (age 15.5) = +18.405 IQ points (normed on age 25)

(1) 18.0 = +0.674 SD (age 15.5); 18.0 becomes 29.0 percentile or −0.553 SD (age 25).

(2) +0.674 minus −0.553 = +1.227 SD or an 18.40 IQ-point gap at that level.

Median (age 15.5) = +15.54 IQ points (normed on age 25)

(1) 14.0 = 0.00 SD (age 15.5); 14.0 becomes 15.0 percentile or −1.036 SD (age 25).

(2) 0.00 minus −1.036 = +1.036 SD or a 15.54 IQ-point gap at that level.

−0.674 SD (age 15.5) = +13.215 (normed on age 25)

(1) 10.0 = −0.674 SD (age 15.5); 10.0 becomes 6.0 percentile or −1.555 SD (age 25).

(2) −0.674 minus −1.555 = +0.881 SD or 13.215 IQ-point gap at that level.

−1.282 SD (age 15.5) = +8.385 (normed on age 25)

(1) 7.0 = −1.282 SD (age 15.5); 7.0 becomes 3.0 percentile or −1.881 SD (age 25).

(2) −1.282 minus −1.881 = +0.559 SD or an 8.385 IQ-point gap at that level.

−1.645 SD (age 15.5) = +7.875 (normed on age 25)

(1) 5.5 = −1.645 (age 15.5); 5.5 becomes 1.5 percentile or −2.170 SD (age 25).

(2) −1.645 minus −2.170 = +0.525 SD or a 7.875 IQ-point gap at that level.

Now subtract the median value (allow for age) from the others

---

Age 15.5 normed on age 25

+1.645 SD = +14.565 IQ points minus +15.54 = *−0.975*
+1.282 SD = +17.34 IQ points minus +15.54 = +1.80
+0.674 SD = +18.405 IQ points minus +15.545 = +2.86
Median = +15.545 IQ points minus +15.545 = NIL
−0.674 SD = +13.215 IQ points minus +15.545 = −2.33
−1.282SD = +8.385 IQ points minus +15.545 = −7.16
−1.645 SD = +7.875 IQ points minus +15.545 = −7.67

The classic pattern holds below the median; at the highest level there is a small negative value. The cumulative values are almost "perfect."

Cumulative totals for age 7.5 normed on age 25

+1.645 SD = +12.59 plus *−0.975* = +11.62
+1.282 SD = +9.63 plus +1.80 = +11.43
+0.674 SD = +7.35 plus +2.86 = +10.21
−0.674 SD = −3.76 plus −2.33 = −6.09
−1.282 SD = (−3.76) plus −7.16 = −10.92
−1.645 SD = (−3.76) plus −7.67 = −11.43

It will be useful, tracking with age, to have 9.5 normed on age 25.

+1.645 SD = +6.22 plus *−0.975* = +5.245
+1.282 SD = +5.48 plus +1.80 = +7.28
+0.674 SD = +3.96 plus +2.86 = +6.82
−0.674 SD = +*0.18* plus −2.33 = −2.15
−1.282 SD = (+*0.18*) plus −7.16 = −6.98
−1.645 SD = (+*0.18*) plus −7.67 = −7.49

And to get values for age 12.5 first by norming it on age 15.5:

+1.645 SD (age 12.5) = +6.75 IQ points (normed on age 15.5)

(1) 41.5 = +1.645 SD (age 12.5); 41.5 becomes 88.39 percentile or +1.195 SD (age 15.5).

(2) +1.645 minus +1.195 = +0.450 SD or a 6.75 IQ-point gap at that level.

+1.282 SD (age 12.5) = +8.34 IQ points (normed on age 15.5)

(1) 39.3 = +1.282 SD (age 12.5); 39.3 becomes 76.61 percentile or +0.7261 SD (age 15.5).

(2) +1.282 minus +0.7261 = +0.556 SD or an 8.34 IQ-point gap at that level.

+0.674 SD (age 12.5) = +7.41 IQ points (normed on age 15.5)

(1) 36.5 = +0.674 SD (age 9.5); 36.5 becomes 57.14 percentile or −0.18 SD (age 15.5).

(2) +0.674 minus −0.180 = +0.494 SD or a 7.41 IQ-point gap at that level.

Median (age 12.5) = +8.23 IQ points (normed on age 15.5)

(1) 33.0 = 0.00 SD (age 12.5); 33.0 becomes 29.17 percentile or −0.5485 SD (age 15.5).

(2) 0.00 minus −0.5485 = +0.5485 SD or an 8.23 IQ-point gap at that level.

−0.674 SD (aged 12.5) = +4.26 points (normed on age 15.5)

(1) 30.5 = −0.674 SD (age 12.5); 30.5 becomes 16.89 percentile or −0.958 SD (age 15.5).

(2) −0.674 minus −0.958 = +0.284 SD or a 4.26 IQ-point gap at that level.

−**1.282 SD** (aged 12.5) = +7.62 IQ points (normed on age 15.5)

(1) 25.9 = −1.282 SD (age 12.5); 25.9 becomes 3.667 percentile or −1.790 SD.

(2) −1.282 minus −1.790 = +0.508 SD or a 7.62 IQ-point gap at that level.

−**1.645 SD**: only a hypothetical estimate at this level − see above.

Now subtract the median value (allow for age) from the others.

Age 12.5 normed on age 15.5
+**1.645 SD** = +6.75 IQ points minus +8.23 = −*1.48*
+**1.282 SD** = +8.34 IQ points minus +8.23 = +0.11
+**0.674 SD** = +7.41 IQ points minus +8.23 = −*0.82*
Median = +8.23 IQ points minus +8.23 = NIL
−**0.674 SD** = +4.26 IQ points minus +8.23 = −3.97
−**1.282 SD** = +7.62 IQ points minus +8.23 = −0.61
−**1.645 SD** = (hypothetical estimate) = −0.61

And to norm 12.5 on age 25
+**1.645 SD** = −*1.48* plus −*0.975* = −2.45
+**1.282 SD** = +0.11 plus +1.80 = +1.91
+**0.674 SD** = −*0.82* plus +2.86 = +2.04
−**0.674 SD** = −3.97 plus −2.33 = −6.30
−**1.282 SD** = −0.61 plus −7.16 = −7.77
−**1.645 SD** = (−0.61) plus −7.67 = −8.38

Coloured Progressive Matrices (2007)

Estimates for young children. The standardization was done close to the most recent standardization of the SPM. Values in bold are relevant to the first norming (4.25/6.25), and the values in brackets relevant to the second (6.25/7.5).

		Age in years						
Percentile	SD	4.25	4.75	5.25	5.75	6.25	6.75	7.5
95	+1.645	**20.50**	24.00	25.50	26.50	(29.00)	30.50	33.00
90	+1.282	**18.86**	22.21	23.79	24.21	(27.79)	28.86	31.79
75	+0.674	**17.00**	18.50	20.50	20.50	(24.50)	26.50	(29.00)
50	—	**15.00**	15.00	17.50	18.00	**(20.50)**	22.50	(26.50)
25	−0.674	**13.00**	13.00	14.50	14.50	**(17.50)**	18.50	(22.50)
10	−1.282	**11.14**	11.14	12.14	12.14	**(13.79)**	14.79	(18.79)
5	−1.645	**10.00**	10.00	11.00	11.00	**(12.00)**	12.50	(16.50)
1	−2.237	8.00	8.00	9.00	9.00	**10.00**	10.00	(13.00)

Note: the calculations below make use of more detailed percentile/raw score equivalents in the CPM table.

+1.645 SD (age 4.25) = +24.675 IQ points (normed on age 6.25)

(1) 20.5 = +1.645 SD (age 4.25); 20.5 becomes 50.00 percentile or nil (age 6.25).

(2) +1.645 minus 0 = +1.645 SD or a +24.675 IQ-point gap at that level.

+1.282 SD (age 4.25) = +24.66 IQ points (normed on age 6.25)

(1) 18.86 = +1.282 SD (age 4.25); 18.86 becomes 35.88 percentile or −0.362 SD (age 6.25).

(2) +1.282 minus −0.362 = +1.644 SD or a 24.66 IQ-point gap at that level.

+0.674 SD (age 4.25) = +21.315 IQ points (normed on age 6.25)

(1) 17.00 = +0.674 SD (age 4.25); 17.0 becomes 22.75 percentile or −0.747 SD (age 6.25).

(2) +0.674 minus −0.747 = +1.421 SD or a 21.315 IQ-point gap at that level.

Median (age 4.25) = +16.035 IQ points (normed on age 6.25)

(1) 15.0 = 0.00 SD (age 4.25); 15.0 becomes 14.25 percentile or −1.069 SD (age 6.25).

(2) 0.00 minus −1.069 = +1.069 SD or a 16.035 IQ-point gap at that level.

−0.674 SD (age 4.25) = +11.325 (normed on age 6.25)

(1) 13.0 = −0.674 SD (age 4.25); 13.0 becomes 7.667 percentile or −1.429 SD (age 6.25).

(2) −0.674 minus −1.429 = +0.755 SD or an 11.325 IQ-point gap at that level.

−1.282 SD (age 4.25) = +9.72 (normed on age 6.25)

(1) 11.14 = −1.282 SD (age 7.5); 11.14 becomes 2.678 percentile or −1.930 SD (age 6.25).

(2) −1.282 minus −1.930 = +0.648 SD or a 9.72 IQ-point gap at that level.

−1.645 SD (age 4.25) = +10.23 (normed on age 6.25)

(1) 10.0 = −1.645 SD (age 4.25); 10.0 becomes 1.0 percentile or −2.327 SD (age 6.25).

(2) −1.645 minus −2.327 = +0.682 SD or a 10.23 IQ-point gap at that level.

Now subtract the median value (allow for age) from the others.

Age 4.25 normed on age 6.25

+1.645 SD = +24.675 IQ points minus +16.035 = +8.64
+1.282 SD = +24.66 IQ points minus +16.035 = +8.625
+0.674 SD = +21.315 IQ points minus +16.035 = +5.28
Median = +16.035 IQ points minus +16.035 = NIL
−0.674 SD = +11.325 IQ points minus +16.035 = −4.71
−1.282 SD = +9.72 IQ points minus +16.035 = −6.315
−1.645 SD = +10.23 IQ points minus +16.035 = −5.805

The classic pattern of family effects asserts itself.

Note: the calculations below make use of more detailed percentile/raw score equivalents in the CPM table.

+1.645 SD (age 6.25) = +14.565 IQ points (normed on age 7.5)

(1) 29.0 = +1.645 SD (age 6.25); 29.0 becomes 75.0 percentile or +0.674 SD (age 7.5).

(2) +1.645 minus +0.674 = +0.971 SD or a +14.565 IQ-point gap at that level.

+1.282 SD (age 6.25) = +14.97 IQ points (normed on age 7.5)

(1) 27.79 = +1.282 SD (age 6.25); 27.79 becomes 61.18 percentile or −0.284 SD (age 7.5).

(2) +1.282 minus −0.284 = +0.998 SD or a 14.97 IQ-point gap at that level.

+0.674 SD (age 6.25) = +15.09 IQ points (normed on age 7.5)

(1) 24.50 = +0.674 SD (age 6.25); 24.50 becomes 37.0 percentile or −0.332 SD (age 7.5).

(2) +0.674 minus −0.332 = +1.006 SD or a +15.09 IQ-point gap at that level.

Median (age 6.25) = +14.925 IQ points (normed on age 7.5)

(1) 20.5 = 0.00 SD (age 6.25); 20.5 becomes 16.0 percentile or −0.995 SD (age 7.5).

(2) 0.00 minus −0.995 = +0.995 SD or a 14.925 IQ-point gap at that level.

−0.674 SD (age 6.25) = +12.03 (normed on age 7.5)

(1) 17.5 = −0.674 SD (age 6.25); 17.5 becomes 7.0 percentile or −1.476 SD (age 7.5).

(2) −0.674 minus −1.476 = +0.802 SD or a 12.03 IQ-point gap at that level.

−1.282 SD (age 6.25) = +12.63 (normed on age 7.5)

(1) 13.79 = −1.282 SD (age 6.25); 13.79 becomes 1.685 percentile or −2.124 SD (age 7.5).

(2) −1.282 minus −2.124 = +0.842 SD or a 12.63 IQ-point gap at that level.

−1.645 SD: no estimate at this level. The raw score for the earlier age is off the scale of the older age. So that various ages will be comparable, I have assumed that this estimate would be identical to the lowest we have (−1.282 SD) and call it a hypothetical estimate.

Now subtract the median value (allow for age) from the others.

Age 6.25 normed on age 7.5

+1.645 SD = +14.565 IQ points minus +14.925 = −0.36
+1.282 SD = +14.97 IQ points minus +14.925 = +0.045
+0.674 SD = +15.09 IQ points minus +14.925 = +0.165
Median = +14.925 IQ points minus +14.925 = NIL
−0.674 SD = +12.03 IQ points minus +14.925 = −2.895
−1.282 SD = +12.63 IQ points minus +14.925 = −2.295
−1.645 SD = (hypothetical estimate) = −2.295

The classic pattern is faint between these years, probably because they are barely one year apart. But age 7.5 is useful because it links with the SPM data. First, of course, when added to ages 4.25/6.25, it gives a cumulative total for the early childhood years, which is near "perfect."

Cumulative totals for age 4.25 normed on age 7.5

+1.645 SD = +8.64 plus −0.36 = +8.28
+1.282 SD = +8.625 plus +0.045 = +8.67
+0.674 SD = +5.28 plus +0.165 = +5.445
−0.674 SD = −4.71 plus −2.895 = −7.605
−1.282 SD = −6.315 plus −2.295 = −8.61
−1.645 SD = −5.805 plus (−2.295) = −8.10

Finally, we have estimates for the full gamut of ages.

Cumulative totals for age 4.25 normed on age 25
+1.645 SD = +8.28 plus +11.62 = +19.90
+1.282 SD = +8.67 plus +11.43 = +20.10
+0.674 SD = +5.445 plus +10.21 = +15.655
−0.674 SD = −7.605 plus −6.09 = −13.695
−1.282 SD = −8.61 plus −10.92 = −19.53
−1.645 SD = −8.10 plus −11.43 = −19.53

Thanks to the cumulative results, we can construct a table tracing family effects with age all the way from age 4.25 to age 25 (see summary in Table AIII1).

Intermediate data (linking 1982 CPM with 1979 SPM with 1992 APM)

Second, I offer the intermediate data.

> **Step I.** Data: raw scores by age at various performance levels.
> **Step II.** How to get the values by SD level: given the great interest of this data, I will detail my calculations throughout.
> **Step III.** Subtract the difference at the median from the difference at all other levels

Standard Progressive Matrices Plus (1979)

First, I norm age 7.5 on 9.5 and then, age 9.5 on 12.5 to get a cumulative total, and then 12.5 on 15.5 to get another (for 7.5 on 15.5). The values in bold are relevant to the first norming (7.5/9.5), the values in (brackets) relevant to the second (9.5/12.5), and the values in [brackets] relevant to the third.

		Age in years								
Percentile	SD	7.5	8.5	9.5	10.5	11.5	12.5	13.5	14.5	15.5
95	+1.645	37	42	(46)	49	51	[53]	54	56	57
90	+1.282	35	40	(44)	47	49	[51]	53	54	[55]
75	+0.674	30	36	(41)	43	45	[(47)]	49	50	[51]
50	—	22	31	(36)	39	41	[(42)]	44	46	[47]
25	−0.674	15	22	(28)	33	36	[(38)]	41	42	[42]
10	−1.282	12	17	19	27	31	[(32)]	35	36	[36]
5	−1.645	11	11	14	22	25	(27)	29	33	[33]

+1.645 SD (age 7.5) = +22.785 IQ points (normed on age 9.5)

(1) 37 = +1.645 SD (age 7.5); 37 becomes 55.0 percentile or +0.126 SD (age 9.5).

(2) +1.645 minus +0.126 = +1.519 SD or a 22.785 IQ-point gap at that level.

+1.282 SD (age 7.5) = +20.40 IQ points (normed on age 9.5)

(1) 35 = +1.282 SD (age 7.5); 35 becomes 46.9 percentile or −0.078 SD (age 9.5).

(2) +1.282 minus −0.078 = +1.360 SD or a 20.40 IQ-point gap at that level.

+0.674 SD (age 7.5) = +17.45 IQ points (normed on age 9.5)

(1) 30 = +0.674 SD (age 7.5); 30 becomes 31.25 percentile or −0.489 SD (age 9.5).

(2) +0.674 minus −0.489 = +1.163 SD or a 17.45 IQ-point gap at that level.

Median (age 7.5) = +15.56 IQ points (normed on age 9.5)

(1) 22 = 0.00 SD (age 7.5); 22 becomes 15.00 percentile or −1.037 SD (age 9.5).

(2) 0.00 minus −1.037 = +1.037 SD or a 15.56 IQ-point gap at that level.

−**0.674 SD** (age 7.5) = +13.22 (normed on age 9.5)

(1) 15 = −0.674 SD (age 7.5); 15 becomes 6.00 percentile or −1.555 SD (age 9.5).

(2) −0.674 minus −1.555 = +0.881 SD or a 13.22 IQ-point gap at that level.

−**1.282 SD** and −**1.645 SD**: no estimates at these levels. Look at the data and you will see that the raw scores for the earlier age are off the scale of the older age. So that various ages will be comparable, I have assumed that the estimates would be identical to the lowest we have (−0.674 SD). I call these hypothetical estimates. Since values tend to rise the further we go below the median, the true values might be a bit higher.

Now subtract the median value (allow for age) from the others.

Age 7.5 normed on age 9.5
+**1.645 SD** = +22.79 IQ points minus +15.56 = +7.23
+**1.282 SD** = +20.40 IQ points minus +15.56 = +4.84
+**0.674 SD** = +17.45 IQ points minus +15.56 = +1.89
Median = +15.56 IQ points minus +15.56 = NIL
−**0.674 SD** = +13.22 IQ points minus +15.56 = −2.34
−**1.282 SD** = (hypothetical estimate) = −2.34
−**1.645 SD** = (hypothetical estimate) = −2.34

The classic pattern of family effects disadvantageous above the median and advantageous below asserts itself!

+**1.645 SD** (age 9.5) = +16.81 IQ points (normed on age 12.5)

(1) 46 = +1.645 SD (age 9.5); 46 becomes 70.00 percentile or +0.5244 SD (age 12.5).

(2) +1.645 minus +0.5244 = +1.12 SD or a 16.81 IQ-point gap at that level.

+1.282 SD (age 9.5) = +15.43 IQ points (normed on age 12.5)

(1) 44 = +1.282 SD (age 9.5); 44 becomes 60.00 percentile or +0.2533 SD (age 12.5).

(2) +1.282 minus +0.2533 = +1.0287 SD or a 15.43 IQ-point gap at that level.

+0.674 SD (age 9.5) = +12.48 IQ points (normed on age 12.5)

(1) 41 = +0.674 SD (age 9.5); 41 becomes 43.75 percentile or −0.158 SD (age 12.5).

(2) +0.674 minus −0.158 = +0.832 SD or a 12.48 IQ-point gap at that level.

Median (age 9.5) = +12.63 IQ points (normed on age 12.5)

(1) 36 = 0.00 SD (age 9.5); 36 becomes 20.00 percentile or −0.842 SD (age 12.5).

(2) 0.00 minus −0.842 = +0.842 SD or a 12.63 IQ-point gap at that level.

−0.674 SD (aged 9.5) = +13.21 points (normed on age 12.5)

(1) 28 = −0.674 SD (age 9.5); 28 becomes 6.00 percentile or −1.555 SD (age 12.5).

(2) −0.674 minus −1.555 = +0.881 SD or a 13.21 IQ-point gap at that level.

−1.282 SD and **−1.645 SD**: only hypothetical estimates at these levels – see above.

Now subtract the median value (allow for age) from the others.

Age 9.5 normed on age 12.5
+1.645 SD = +16.81 IQ points minus +12.63 = +4.18
+1.282 SD = +15.43 IQ points minus +12.63 = +2.80
+0.674 SD = +12.48 IQ points minus +12.63 = −0.15
Median = +12.63 IQ points minus +12.63 = NIL
−0.674 SD = +13.21 IQ points minus +12.63 = +0.58
−1.282 SD = (hypothetical estimate) = +0.58
−1.645 SD = (hypothetical estimate) = +0.58

The classic pattern holds only at the upper levels with slightly negative effects elsewhere. Nonetheless, the cumulative values are "correct," though still low below the median.

Cumulative totals for age 7.5 normed on age 12.5
+1.645 SD = +7.23 plus +4.18 = +11.41
+1.282 SD = +4.84 plus +2.80 = +7.64
+0.674 SD = +1.89 plus −0.15 = +1.74
−0.674 SD = −2.34 plus +0.58 = −1.76
−1.282 SD = (−2.34) plus (+0.58) = −1.76
−1.645 SD = (−2.34) plus (+0.58) = −1.76

+1.645 SD (age 12.5) = +10.66 IQ points (normed on age 15.5)

(1) 53 = +1.645 SD (age 12.5); 53 becomes 82.5 percentile or +0.9346 SD (age 15.5).

(2) +1.645 minus +0.9346 = +0.7104 SD or a 10.66 IQ-point gap at that level.

+1.282 SD (age 12.5) = +9.12 IQ points (normed on age 15.5)

(1) $51 = +1.282$ SD (age 12.5); 51 becomes 75.00 percentile or $+0.674$ SD (age 15.5).

(2) $+1.282$ minus $+0.674 = +0.608$ SD or a 9.12 IQ-point gap at that level.

+0.674 SD (age 12.5) $= +10.11$ IQ points (normed on age 15.5)

(1) $47 = +0.674$ SD (age 12.5); 47 becomes 50.00 percentile or -0.00 SD (age 15.5).

(2) $+0.674$ minus $-0.00 = +0.674$ SD or a 10.11 IQ-point gap at that level.

Median (age 12.5) $= +10.11$ IQ points (normed on age 15.5)

(1) $42 = 0.00$ SD (age 12.5); 42 becomes 25.00 percentile or -0.674 SD (age 15.5).

(2) 0.00 minus $-0.674 = +0.674$ SD or a 10.11 IQ-point gap at that level.

−0.674 SD (aged 12.5) $= +3.91$ points (normed on age 15.5)

(1) $38 = -0.674$ SD (age 12.5); 38 becomes 17.50 percentile or -0.9346 SD (age 15.5).

(2) -0.674 minus $-0.9346 = +0.2606$ SD or a 3.91 IQ-point gap at that level.

−1.282 SD (aged 12.5) $= +8.30$ points (normed on age 15.5)

(1) $32 = -1.282$ SD (age 12.5); 32 becomes 3.33 percentile or -1.835 SD (age 15.5).

(2) -1.282 minus $-1.835 = +0.553$ SD or a 8.30 IQ-point gap at that level.

−1.645 SD: only hypothetical estimate at this level – see above.

Now subtract the median value (allow for age) from the others.

Age 12.5 normed on age 15.5
+1.645 SD = +10.66 IQ points minus +10.11 = +0.55
+1.282 SD = +9.12 IQ points minus +10.11 = −0.99
+0.674 SD = +10.11 IQ points minus +10.11 = NIL
Median = +10.11 IQ points minus +10.11 = NIL
−0.674 SD = +3.91 IQ points minus +10.11 = −6.20
−1.282 SD = +8.30 IQ points minus +10.11 = −1.71
−1.645 SD = (hypothetical estimate) = −1.71

Family effects above the median are mainly random but advantages below are significant. Nonetheless, the cumulative values are largely "correct" at both age 15.5 and 25.

Cumulative totals for age 7.5 normed on age 15.5
+1.645 SD = +11.41 plus +0.55 = +11.96
+1.282 SD = +7.64 plus −0.99 = +.65
+0.674 SD = +1.74 plus 0.00 = +1.74
−0.674 SD = −1.76 plus −6.20 = −7.96
−1.282 SD = (−1.76) plus −1.71 = (−3.47)
−1.645 SD = (−1.76) plus (−1.71) = (−3.47)

Advanced Progressive Matrices (1992)

Note: this standardization is used here as it was in the latest data above. It dates from between the latest and the intermediate years and should be valid to the same degree for both. Recall that I make ages 18–32 (median 25) the target ages because performance peaks then (without variation) and declines thereafter. Below I merely use its results to get an overall cumulative total for ages 7.5 normed on age 25.

Cumulative totals for age 7.5 normed on age 25

+1.645 SD = +11.96 plus −*0.975* = +10.99
+1.282 SD = +6.65 plus +1.80 = +8.45
+0.674 SD = +1.74 plus +2.86 = +4.60
−0.674 SD = −7.96 plus −2.33 = −10.29
−1.282 SD = (−3.47) plus −7.16 = −10.63
−1.645 SD = (−3.47) plus −7.67 = −10.63

It will be useful, tracking with age, to have age 9.5 normed on age 25. This means adding 9.5 normed on 12.5, to 12.5 normed on 15.5, and to 15.5 normed on 25:

+1.645 SD = +4.18 plus +0.55 plus −*0.975* = +3.76
+1.282 SD = +2.80 plus −*0.99* plus +1.80 = +3.61
+0.674 SD = −*0.15* plus 0.00 plus +2.86 = +2.71
−0.674 SD = +*0.58* plus −6.20 plus −2.33 = −2.15
−1.282 SD = (+*0.58*) plus −1.71 plus −7.16 = −8.29
−1.645 SD = (+*0.58*) plus −1.71 plus −7.67 = −8.80

It will be useful, tracking with age, to have age 12.5 normed on age 25. This means adding 12.5 normed on 15.5 to 15.5 normed on 25.

+1.645 SD = +0.55 plus −*0.975* = −*0.425*
+1.282 SD = −*0.99* plus +1.80 = +0.81
+0.674 SD = 0.00 plus +2.86 = +2.86
−0.674 SD = −6.20 plus −2.33 = −8.53
−1.282 SD = −1.71 plus −7.16 = −8.87
−1.645 SD = (−1.71) plus −7.67 = −9.38

Coloured Progressive Matrices (1982)

Estimates for young children. The standardization was done close to the 1979 standardization of the SPM. The earliest age is 5.5. To norm 5.5 on 7.5 requires two steps: norming 5.5 on age 6.25 and norming 6.25 on age 7.5. The values in bold are relevant to the first and those in brackets to the second.

		Age in years					
Percentile	SD	5.50	6.00	6.25	6.50	7.00	7.50
95	+1.645	**22**	24	(25.0)	26	28	31
90	+1.282	**20**	21	**(22.0)**	23	25	(28)
75	+0.674	**18**	19	**(19.5)**	20	21	(23)
50	—	**15**	16	**(16.5)**	17	18	(20)
25	−0.674	**12**	13	**(13.5)**	14	16	(17)
10	−1.282	**10**	11	**11.5**	12	13	(14)
5	−1.645	8	9	**10.0**	11	12	(13)

+1.645 SD (age 5.5) = +5.445 IQ points (normed on age 6.25)

(1) 22 = +1.645 SD (age 5.5); 22 becomes 90.0 percentile or +1.282 SD (age 6.25).

(2) +1.645 minus +1.282 = 0.363 SD or a +5.445 IQ-point gap at that level.

+1.282 SD (age 5.5) = +7.65 IQ points (normed on age 6.25)

(1) 20 = +1.282 SD (age 5.5); 20 becomes 78.0 percentile or +0.772 SD (age 6.25).

(2) +1.282 minus +0.772 = +0.510 SD or a 7.65 IQ-point gap at that level.

+0.674 SD (age 5.5) = +5.325 IQ points (normed on age 6.25)

(1) 18 = +0.674 SD (age 5.5); 18 becomes 62.5 percentile or +0.319 SD (age 6.25).

(2) +0.674 minus +0.319 = +0.355 SD or a +5.325 IQ-point gap at that level.

Median (age 5.5) = +4.785 IQ points (normed on age 6.25)

(1) 15 = 0.00 SD (age 5.5); 15 becomes 37.5 percentile or −0.319 SD (age 6.25).

(2) 0.00 minus −0.319 = +0.319 SD or a 4.785 IQ-point gap at that level.

−0.674 SD (age 5.5) = +6.27 (normed on age 6.25)

(1) 12 = −0.674 SD (age 5.5); 12 becomes 13.75 percentile or −1.092 SD (age 6.25).

(2) −0.674 minus −1.092 = +0.418 SD or a 6.27 IQ-point gap at that level.

−1.282 SD (age 5.5) = +5.445 (normed on age 6.25)

(1) 10 = −1.282 SD (age 5.5); 10 becomes 5.00 percentile or −1.645 SD (age 6.25).

(2) −1.282 minus −1.645 = +0.363 SD or a 5.445 IQ-point gap at that level.

−1.645 SD: no estimates at these levels. The raw score for the earlier age is off the scale of the older age. So that various ages will be comparable, I have assumed that this estimate would be identical to the lowest we have (−1.282 SD) and call it a hypothetical estimate.

Now subtract the median value (allow for age) from the others.

Age 5.5 normed on age 6.25

+1.645 SD = +5.45 IQ points minus +4.79 = +0.66
+1.282 SD = +7.65 IQ points minus +4.79 = +2.86
+0.674 SD = +5.33 IQ points minus +4.79 = +0.54
Median = +4.79 IQ points minus +4.79 = NIL
−0.674 SD = +6.27 IQ points minus +4.79 = *+1.48*
−1.282 SD = +5.445 IQ points minus +4.79 = *+0.665*
−1.645 SD = (hypothetical estimate) = *+0.665*

Between these ages, family effects are almost random.

+1.645 SD (age 6.25) = +11.505 IQ points (normed on age 7.5)

(1) 25 = +1.645 SD (age 6.25); 25 becomes 81.0 percentile or +0.878 SD (age 7.5).

(2) +1.645 minus 0.878 = 0.767 SD or an 11.505 IQ-point gap at that level.

+1.282 SD (age 6.25) = +12.765 IQ points (normed on age 7.5)

(1) 22 = +1.282 SD (age 6.25); 22 becomes 66.67 percentile or +0.431 SD (age 7.5).

(2) +1.282 minus +0.431 = +0.851 SD or a 12.765 IQ-point gap at that level.

+0.674 SD (age 6.25) = +11.685 IQ points (normed on age 7.5)

(1) 19.5 = 0.674 SD (age 6.25); 19.5 becomes 45.83 percentile or −0.105 SD (age 7.5).

(2) +0.674 minus −0.105 = +0.779 SD or an +11.685 IQ-point gap at that level.

Median (age 6.25) = +12.63 IQ points (normed on age 7.5)

(1) 16.5 = 0.00 SD (age 6.25); 16.5 becomes 20.0 percentile or −0.842 SD (age 7.5).

(2) 0.00 minus −0.842 = +0.842 SD or a 12.63 IQ-point gap at that level.

−0.674 SD (age 6.25) = +12.375 (normed on age 7.5)

(1) 13.5 = −0.674 SD (age 6.25); 13.5 becomes 7.5 percentile or −1.499 SD (age 7.5).

(2) −0.674 minus −1.499 = +0.825 SD or a 12.375 IQ-point gap at that level.

−1.282 SD and **−1.645 SD**: no estimates at these levels. The raw score for the earlier age is off the scale of the older age. So that various ages will be comparable, I have assumed that this estimate would be identical to the lowest we have (−0.674 SD) and call it a hypothetical estimate.

Now subtract the median value (allow for age) from the others.

Age 6.25 normed on age 7.5
+1.645 SD = +11.51 IQ points minus +12.63 = *−1.12*
+1.282 SD = +12.77 IQ points minus +12.63 = +0.14
+0.674 SD = +11.69 IQ points minus +12.63 = *−0.94*
Median = +12.63 IQ points minus +12.63 = NIL
−0.674 SD = +12.34 IQ points minus +12.63 = −0.29
−1.282 SD = (hypothetical estimate) = −0.29
−1.645 SD = (hypothetical estimate) = −0.29

This confirms that at the preschool level, there are no additional family effects to those that show at older ages.

Finally, we have estimates for the full gamut of ages. This means adding 5.5 normed on 6.25, to 6.25 normed on 7.5, and to 7.5 normed on 25.

Cumulative totals for age 5.5 normed on age 25
+1.645 SD = +0.66 plus *−1.12* plus +10.99 = +10.53
+1.282 SD = +2.86 plus +0.14 plus +8.45 = +11.45
+0.674 SD = +0.54 plus *−0.94* plus +4.60 = +4.20
−0.674 SD = *+1.48* plus −0.29 plus −10.29 = −9.10
−1.282 SD = *+0.67* plus (−0.29) plus −10.63 = −10.25
−1.645 SD = (*+0.67*) plus (−0.29) plus −10.63 = −10.25

Thanks to the cumulative results, we can construct a table tracing family effects with age all the way from age 5.50 to age 25 (see the summary in Table AIII1).

Early data (linking 1949 CPM with 1938 SPM-children with 1942 SPM-adults)

Finally, I offer the early data.

> **Step I.** Data: raw scores by age at various performance levels.

Step II. How to get the values by SD level: given the great interest of this data, I will detail my calculations throughout.

Step III. Subtract the difference at the median from the difference at all other levels

Standard Progressive Matrices (1938 and 1942)

There is no APM data from these early years but the SPM was normed on children in 1938 and adults in 1942. The manual treats these as one continuous series of scores beginning with age 8 and ending with adults, by which it means their peak ages of performance – namely, ages 20 to 25 (or 22.5). The relevant data are given below. We will link age 8 normed on age 9.5, with age 9.5 normed on age 12.5, with age 12.5 normed on age 14 (the oldest child age), with age 14 normed on age 22.5.

		Age in years								
Percentile	**SD**	**8**	8.5	**9.5**	10.5	11.5	**12.5**	13.5	**14**	**22.5**
95	+1.645	38	39	44	48	51	51	53	53	55
90	+1.282	34	36	41	45	49	50	51	52	54
75	+0.674	24	29	34	40	43	46	48	48	49
50	—	18	21	28	33	37	41	44	44	44
25	−0.674	13	15	18	23	29	34	37	38	37
10	−1.282	12	12	13	15	18	26	28	28	28
5	−1.645	10	11	11	13	15	17	21	23	23

+1.645 SD (age 8) = +10.02 IQ points (normed on age 9.5)

(1) 38 = +1.645 SD (age 8); 38 becomes 83.57 percentile or +0.977 SD (age 9.5).

(2) +1.645 minus +0.977 = +0.668 SD or a 10.02 IQ-point gap at that level.

+1.282 SD (age 8) = +9.12 IQ points (normed on age 9.5)

(1) 34 = +1.282 SD (age 8); 34 becomes 75.0 percentile or +0.674 SD (age 9.5).

(2) +1.282 minus +0.674 = +0.608 SD or a 9.12 IQ-point gap at that level.

+0.674 SD (age 8) = +13.905 IQ points (normed on age 9.5)

(1) 24 = +0.674 SD (age 8); 24 becomes 40.0 percentile or −0.253 SD (age 9.5).

(2) +0.674 minus −0.253 = +0.927 SD or a 13.905 IQ-point gap at that level.

Median (age 8) = +10.11 IQ points (normed on age 9.5)

(1) 18 = 0.00 SD (age 8); 18 becomes 25.00 percentile or −0.674 SD (age 9.5).

(2) 0.00 minus −0.674 = +0.674 SD or a 10.11 IQ-point gap at that level.

−0.674 SD (age 8) = +9.12 (normed on age 9.5)

(1) 13 = −0.674 SD (age 8); 13 becomes 10.00 percentile or −1.282 SD (age 9.5).

(2) −0.674 minus −1.282 = +0.608 SD or a 9.12 IQ-point gap at that level.

−1.282 SD (age 8) = +2.355 (normed on age 9.5)

(1) 12 = −1.282 SD (age 8); 12 becomes 7.50 percentile or −1.439 SD (age 9.5).

(2) 1.282 minus −1.439 = +0.157 SD or a 2.355 IQ-point gap at that level.

−1.645 SD: no estimate at this level. The raw score for the earlier age is off the scale of the older age. So that various ages will be comparable, I have assumed that this estimate would be identical to the lowest we have (−1.282 SD) and call it a hypothetical estimate.

Now subtract the median value (allow for age) from the others.

Age 8 normed on age 9.5
+1.645 SD = +10.02 IQ points minus +10.11 = *−0.09*
+1.282 SD = +9.12 IQ points minus +10.11 = *−0.99*
+0.674 SD = +13.91 IQ points minus +10.11 = *+2.80*
Median = +10.11 IQ points minus +10.11 = NIL
−0.674 SD = +9.12 IQ points minus +10.11 = −0.99
−1.282 SD = +2.36 IQ points minus +10.11 = −8.64
−1.645 SD = (hypothetical estimate) = −8.64

The high percentiles (90th and 95th) are unique in showing no family effects, and the latter are large only at the low percentiles (10th and 5th).

+1.645 SD (age 9.5) = +18.90 IQ points (normed on age 12.5)

(1) 44 = +1.645 SD (age 9.5); 44 becomes 65.0 percentile or +0.385 SD (age 12.5).

(2) +1.645 minus +0.385 = 1.260 SD or an 18.90 IQ-point gap at that level.

+1.282 SD (age 9.5) = +19.23 IQ points (normed on age 12.5)

(1) 41 = +1.282 SD (age 9.5); 41 becomes 50.0 percentile or 0.000 SD (age 12.5).

(2) +1.282 minus 0.000 = +1.282 SD or a 19.23 IQ-point gap at that level.

+0.674 SD (age 9.5) = +20.22 IQ points (normed on age 12.5)

(1) 34 = +0.674 SD (age 9.5); 34 becomes 13.75 percentile or −1.115 SD (age 12.5).

(2) +0.674 minus −0.674 = +1.348 SD or a 20.22 IQ-point gap at that level.

Median (age 9.5) = +16.38 IQ points (normed on age 12.5)

(1) 28 = 0.00 SD (age 9.5); 28 becomes 13.75 percentile or −1.092 SD (age 12.5).

(2) 0.00 minus −1.092 = +1.092 SD or a 16.38 IQ-point gap at that level.

−0.674 SD (age 9.5) = +13.79 (normed on age 12.5)

(1) 18 = −0.674 SD (age 9.5); 18 becomes 5.556 percentile or −1.593 SD (age 12.5).

(2) −0.674 minus −1.593 = +0.919 SD or a 13.785 IQ-point gap at that level.

−1.282 SD and **−1.645 SD**: no estimates at these levels. The raw score for the earlier age is off the scale of the older age. So that various ages will be comparable, I have assumed that these estimates would be identical to the lowest we have (−0.674 SD) and call them hypothetical estimates.

Now subtract the median value (allow for age) from the others.

Age 9.5 normed on age 12.5
+1.645 SD = +18.90 IQ points minus +16.38 = +2.52
+1.282 SD = +19.23 IQ points minus +16.38 = +2.85
+0.674 SD = +20.22 IQ points minus +16.38 = +3.84
Median = +16.38 IQ points minus +16.38 = NIL
−0.674 SD = +13.79 IQ points minus +16.38 = −2.59
−1.282 SD = (hypothetical estimate) = −2.59
−1.645 SD = (hypothetical estimate) = −2.59

This is pretty much the classic pattern.

+1.645 SD (age 12.5) = +8.30 IQ points (normed on age 14)

(1) 51 = +1.645 SD (age 12.5); 51 becomes 86.25 percentile or +1.092 SD (age 14).

(2) +1.645 minus +1.092 = +0.553 SD or an 8.295 IQ-point gap at that level.

+1.282 SD (age 12.5) = +5.21 IQ points (normed on age 14)

(1) 50 = +1.282 SD (age 12.5); 50 becomes 82.5 percentile or +0.9346 SD (age 14).

(2) +1.282 minus +0.9346 = +0.3474 SD or a 5.21 IQ-point gap at that level.

+0.674 SD (age 12.5) = +5.33 IQ points (normed on age 14)

(1) 46 = +0.674 SD (age 12.5); 46 becomes 62.5 percentile or +0.3187 SD (age 14).

(2) +0.674 minus +0.3187 = +0.3553 SD or a 5.33 IQ-point gap at that level.

Median (age 12.5) = +4.78 IQ points (normed on age 14)

(1) 41 = 0.00 SD (age 12.5); 41 becomes 37.5 percentile or −0.3187 SD (age 14).

(2) 0.00 minus −0.3187 = +0.3187 SD or a 4.78 IQ-point gap at that level.

−0.674 SD (age 12.5) = +3.06 (normed on age 14)

(1) 34 = −0.674 SD (age 12.5); 34 becomes 19.0 percentile or −0.8779 SD (age 14).

(2) −0.674 minus −0.8779 = +0.2039 SD or a 3.06 IQ-point gap at that level.

−1.282 SD (age 12.5) = +1.85 (normed on age 14)

(1) 26 = −1.282 SD (age 12.5); 26 becomes 8.0 percentile or −1.4053 SD (age 14).

(2) −1.282 minus −1.4053 = +0.1233 SD or a 1.85 IQ-point gap at that level.

−**1.645 SD**: no estimate at this level. The raw score for the earlier age is off the scale of the older age. So that various ages will be comparable, I have assumed that this estimate would be identical to the lowest we have (−1.282 SD) and call it a hypothetical estimate.

Now subtract the median value (allow for age) from the others.

Age 12.5 normed on age 14.0

+1.645 SD = +8.30 IQ points minus +4.78 = +3.52
+1.282 SD = +5.21 IQ points minus +4.78 = +0.43
+0.674 SD = +5.33 IQ points minus +4.78 = +0.55
Median = +4.78 IQ points minus +4.78 = NIL
−0.674 SD = +3.06 IQ points minus +4.78 = −1.72
−1.282 SD = +1.85 IQ points minus +4.78 = −2.93
−1.645 SD = (hypothetical estimate) = −2.93

This is almost exactly the classic pattern.

+1.645 SD (age 14) = +7.78 IQ points (normed on age 22.5)

(1) 53 = +1.645 SD (age 14); 53 becomes 87.0 percentile or +1.1264 SD (age 22.5).

(2) +1.645 minus +1.1264 = +0.5186 SD or a 7.78 IQ-point gap at that level.

+1.282 SD (age 14) = +4.31 IQ points (normed on age 22.5)

(1) 52 = +1.282 SD (age 14); 52 becomes 84.0 percentile or +0.9946 SD (age 22.5).

(2) +1.282 minus +0.9946 = +0.2874 SD or a 4.31 IQ-point gap at that level.

+0.674 SD (age 14) = +2.24 IQ points (normed on age 22.5)

(1) 48 = +0.674 SD (age 14); 48 becomes 70.0 percentile or +0.5244 SD (age 22.5).
(2) +0.674 minus +0.5244 = +0.1496 SD or a 2.24 IQ-point gap at that level.

Median (age 14) = 0.00 IQ points (normed on age 22.5)

(1) 44 = 0.00 SD (age 14); 44 becomes 50.0 percentile or 0.00 SD (age 22.5).
(2) 0.00 minus 0.00 = 0.00 SD or a NIL point gap at that level.

−0.674 SD (age 14) = −2.24 (normed on age 22.5)

(1) 38 = −0.674 SD (age 14); 38 becomes 30.0 percentile or −0.5244 SD (age 22.5).
(2) −0.674 minus −0.5244 = −0.1496 SD or a 2.24 IQ-point gap at that level.

−1.282 SD (age 14) = 0.00 (normed on age 22.5)

(1) 28 = −1.282 SD (age 14); 28 becomes 10.0 percentile or −1.282 SD (age 22.5).
(2) −1.282 minus −1.282 = 0.00 SD or a NIL point gap at that level.

−1.645 SD (age 14) = 0.00 (normed on age 22.5)

(1) 23 = −1.645 SD (age 14); 23 becomes 5.0 percentile or −1.645 SD (age 22.5).
(2) −1.645 minus −1.645 = 0.00 SD or a NIL point gap at that level.

Now subtract the median value (allow for age) from the others for age 14 normed on age 22.5. But since the result is NIL at the median, all of the values stand as above.

+1.645 SD = +7.78
+1.282 SD = +4.31
+0.674 SD = +2.24
Median = NIL
−0.674 SD = −2.24
−1.282 SD = 0.00
−1.645 SD = 0.00

Shows further family effects above the median but little or none below.

Coloured Progressive Matrices (1949)

Estimates for young children. The lowest age is 5.5 and it makes sense to norm it on age 8 as the age that provides a link to the SPM. To norm 5.5 on 8 requires two steps: norming 5.5 on age 7 and norming 7 on age 8.

		Age in years					
Percentile	**SD**	**5.50**	**6.00**	**6.50**	**7.00**	**7.50**	**8.00**
95	+1.645	19	21	23	24	25	26
90	+1.282	17	20	21	22	23	24
75	+0.674	15	17	18	19	20	21
50	—	14	15	15	16	17	18
25	−0.674	12	13	14	14	15	16
10	−1.282	—	12	12	13	14	14
5	−1.645	—	—	—	12	12	13

+1.645 SD (age 5.5) = +14.57 IQ points (normed on age 7)

(1) 19 = +1.645 SD (age 5.5); 19 becomes 75.0 percentile or +0.674 SD (age 7).

(2) +1.645 minus +0.674 = +0.971 SD or a 14.565 IQ-point gap at that level.

+1.282 SD (age 5.5) = +16.08 IQ points (normed on age 7)

(1) 17 = +1.282 SD (age 5.5); 17 becomes 58.33 percentile or +0.210 SD (age 7).

(2) +1.282 minus +0.210 = +1.072 SD or a 16.08 IQ-point gap at that level.

+0.674 SD (age 5.5) = +14.89 IQ points (normed on age 7)

(1) 15 = +0.674 SD (age 5.5); 15 becomes 37.5 percentile or −0.3187 SD (age 7).

(2) +0.674 minus −0.3187 = +0.9927 SD or a 14.89 IQ-point gap at that level.

Median (age 5.5) = +10.11 IQ points (normed on age 7)

(1) 14 = 0.00 SD (age 5.5); 14 becomes 25.0 percentile or −0.674 SD (age 7).

(2) 0.00 minus −0.674 = +0.674 SD or a 10.11 IQ-point gap at that level.

−0.674 SD (age 5.5) = +14.57 (normed on age 7)

(1) 12 = −0.674 SD (age 5.5); 12 becomes 5.0 percentile or −1.645 SD (age 7).

(2) −0.674 minus −1.645 = +0.971 SD or a 14.565 IQ-point gap at that level.

−1.282 SD and **−1.645 SD**: no estimates at these levels because no raw scores for age 5.5 are provided. So that various ages will be comparable, I have assumed that these estimates would be identical to the lowest we have (−0.674 SD) and call them hypothetical estimates.

Now subtract the median value (allow for age) from the others.

Age 5.5 normed on age 7
+1.645 SD = +14.57 IQ points minus +10.11 = +4.46
+1.282 SD = +16.08 IQ points minus +10.11 = +5.97
+0.674 SD = +14.89 IQ points minus +10.11 = +4.78
Median = +10.11 IQ points minus +10.11 = NIL
−0.674 SD = +14.57 IQ points minus +10.11 = +4.46
−1.282 SD = (hypothetical estimate) = +4.46
−1.645 SD = (hypothetical estimate) = +4.46

This shows the expected effects above the median, but the one result below is contrary.

+1.645 SD (age 7) = +5.45 IQ points (normed on age 8)

(1) 24 = +1.645 SD (age 7); 24 becomes 90.0 percentile or +1.282 SD (age 8).

(2) +1.645 minus +1.282 = +0.363 SD or a 5.445 IQ-point gap at that level.

+1.282 SD (age 7) = +6.60 IQ points (normed on age 8)

(1) 22 = +1.282 SD (age 7); 22 becomes 80.0 percentile or +0.8418 SD (age 8).

(2) +1.282 minus +0.8418 = +0. 4402 SD or a 6.60 IQ-point gap at that level.

+0.674 SD (age 7) = +6.96 IQ points (normed on age 8)

(1) 19 = +0.674 SD (age 7); 19 becomes 58.33 percentile or +0.210 SD (age 8).

(2) +0.674 minus −0.210 = +0.464 SD or a 6.96 IQ-point gap at that level.

Median (age 7) = +10.11 IQ points (normed on age 8)

(1) 16 = 0.00 SD (age 7); 16 becomes 25.0 percentile or −0.674 SD (age 8).

(2) 0.00 minus −0.674 = +0.674 SD or a 10.11 IQ-point gap at that level.

−0.674 SD (age 7) = +9.12 (normed on age 8)

(1) 14 = −0.674 SD (age 7); 14 becomes 10.0 percentile or −1.2817 SD (age 8).

(2) −0.674 minus −1.2817 = +0.6077 SD or a 9.12 IQ-point gap at that level.

−1.282 SD (age 7) = +5.45 (normed on age 8)

(1) 13 = −1.282 SD (age 7); 12 becomes 5.0 percentile or −1.645 SD (age 8).

(2) −1.282 minus −1.645 = +0.363 SD or a 5.445 IQ-point gap at that level.

−1.645 SD: no estimate at this level. The raw score for the earlier age is off the scale of the older age. So that various ages will be comparable, I have assumed that this estimate would be identical to the lowest we have (−1.282 SD) and call it a hypothetical estimate.

Now subtract the median value (allow for age) from the others.

Age 7 normed on age 8
+1.645 SD = +5.45 IQ points minus +10.11 = *−4.66*
+1.282 SD = +6.60 IQ points minus +10.11 = *−3.51*
+0.674 SD = +6.96 IQ points minus +10.11 = *−3.15*
Median = +10.11 IQ points minus +10.11 = NIL
−0.674 SD = +9.12 IQ points minus +10.11 = *−0.99*
−1.282 SD = +5.45 IQ points minus +10.11 = *−4.66*
−1.645 SD = (hypothetical estimate) = *−4.66*

This shows the expected effects below the median, but the results above are contrary. In other words, it is the reverse of the data from age 5.5 normed on age 7. It is only fair to note that the early standardization of the CPM was based on 608 schoolchildren from Dumfries in Scotland, which comes to about 51 for each of the twelve ages in the complete table; and that some of the raw scores from all the ages I have used are described as "extrapolated for smooth working." When the cumulative values from the CPM are calculated below, the overall picture is that there were no *additional* family effects from the preschool years. This does not mean there were no family effects, simply that they were constant (and sizable) from ages 5.5 to 8.

Cumulative total for age 5.5 normed on age 8
+1.645 SD = +4.46 plus −4.66 = −0.20
+1.282 SD = +5.97 plus −3.51 = +2.46
+0.674 SD = +4.78 plus −3.15 = +2.63
−0.674 SD = +4.66 plus −0.99 = +3.67
−1.282 SD = (+4.66) plus −4.66 = 0.00
−1.645 SD = (+4.66) plus (−4.66) = 0.00

We can now trace family effects with age all the way from age 5.5 to age 22.5 (the values calculated below are in Table AIII1).

Beginning with age 5.5 normed on age 22.5, this means adding 5.5 on 8, to 8 on 9.5, to 9.5 on 12.5, to 12.5 on 14.0, to 14.0 on 22.5.

+1.645 SD = −0.20 plus −0.09 plus +2.52 plus +3.52 plus +7.78 = +13.53
+1.282 SD = +2.46 plus −0.99 plus +2.85 plus +0.43 plus +4.31 = +9.06
+0.674 SD = +2.63 plus +2.80 plus +3.84 plus +0.55 plus +2.24 = +12.06
−0.674 SD = +3.67 plus −0.99 plus −2.59 plus −1.72 plus −2.24 = −3.87
−1.282 SD = 0.00 plus −8.64 plus −2.59 plus −2.93 plus 0.00 = −14.16
−1.645 SD = 0.00 plus −8.64 plus −2.59 plus −2.93 plus 0.00 = (−4.16)

Age 8 normed on age 22.5 means subtracting 5.5 to 8 from the above total values.

+1.645 SD = +13.53 minus −*0.20* = +13.73
+1.282 SD = +9.06 minus +2.46 = +6.60
+0.674 SD = +12.06 minus +2.63 = +9.43
−0.674 SD = −3.87 mimus +*3.67* = −7.54
−1.282 SD = −14.16 minus 0.00 = −14.16
−1.645 SD = −14.16 minus 0.00 = −14.16

Age 9.50 normed on age 22.5 means subtracting 8–9.50 from the above total values.

+1.645 SD = +13.73 minus −*0.09* = +3.82
+1.282 SD = +6.60 minus −*0.99* = +7.59
+0.674 SD = +9.43 minus +2.80 = +6.63
−0.674 SD = −7.54 minus −0.99 = −6.55
−1.282 SD = −14.16 minus −8.64 = −5.52
−1.645 SD = −14.16 minus −8.64 = −5.52

Age 12.50 normed on age 22.5 means subtracting 9.50–12.50 from the above total values.

+1.645 SD = +13.82 minus +2.52 = +11.30
+1.282 SD = +7.59 minus +2.85 = +4.74
+0.674 SD = +6.63 minus +3.84 = +2.79
−0.674 SD = −6.55 minus −2.59 = −3.96
−1.282 SD = −5.52 minus −2.59 = −2.93
−1.645 SD = −5.52 minus −2.59 = −2.93

Age 14 has already been normed on age 22.5 - the results were:

+1.645 SD = +7.78
+1.282 SD = +4.31
+0.674 SD = +2.24
Median = NIL
−0.674 SD = −2.24
−1.282 SD = 0.00
−1.645 SD = 0.00

The summary table (Table AIII1) presents the results from all three sets of standardizations.

Table AIII1 Family effect by age on Raven's at three times (summary table)

The 1st and 2nd sets of standardizations are normed on the target age of 18–32 (25); the 3rd is normed on the target age of 20–25 (22.5)

Percentile	+/– SD	4.25	5.50	7.50	8.00	9.50	12.50	14.00	15.50
		Results from the 2007/2008/1992 standardizations							
95	+1.645	+19.90		+11.62		+5.25	−2.45		−0.975
90	+1.282	+20.10		+11.43		+7.28	+1.91		+1.80
75	+0.645	+15.66		+10.21		+6.82	+2.04		+2.86
25	−0.645	−13.70		−6.09		−2.15	−6.30		−2.33
10	−1.282	−19.53		−10.92		−6.98	−7.77		−7.16
5	−1.645	−19.53		−11.43		−7.49	−8.38		−7.67
		Results from the 1982/1979/1992 standardizations							
95	+1.645		+10.53	+10.99		+3.76	−0.425		−0.975
90	+1.282		+11.45	+8.45		+3.61	+0.81		+1.80
75	+0.645		+4.20	+4.60		+2.71	+2.86		+2.86
25	−0.645		−9.10	−10.29		−2.15	−8.53		−2.33
10	−1.282		−10.25	−10.63		−8.29	−8.87		−7.16
5	−1.645		−10.25	−10.63		−8.80	−9.38		−7.67

Results from the 1949/1938/1942 standardizations

95	+1.645	+13.53	+13.73	+13.82	+11.30	+7.78
90	+1.282	+9.06	+6.60	+7.59	+4.74	+4.41
75	+0.645	+12.06	+9.43	+6.63	+2.79	+2.24
25	−0.645	−3.87	−7.54	−6.55	−3.96	−2.24
10	−1.282	−14.16	−14.16	−5.52	−2.93	0.00
5	−1.645	−14.16	−14.16	−5.52	−2.93	0.00

243

Table AIII2 Comparison from the most recent Raven's and Wechsler Vocabulary data

Raven's: all ages normed on the target ages of 18–32 (25)							
Results from the 2007/2008/1992 standardizations							
%tile	4.25	7.50	9.50	12.50	15.50	17.5	18	20–24
95	+19.90	+11.62	+5.25	−2.45	−0.975	—	—	—
82.5	+17.88	+10.82	+7.05	+1.98	+2.33	—	—	—
17.5	−16.62	−8.51	−4.57	−7.04	−4.75	—	—	—
5	−19.53	−11.43	−7.49	−8.38	−7.67	—	—	—
Wechsler Vocabulary: all ages normed on target ages of 45–54							
Results from the 2002/2002/2007 standardizations							
%tile	4.00	6.75	9.25	11.50	14.50	17.5	18	20–24
98	+19.72	+8.91	+4.45	+1.91	+0.75	+0.25	−2.25	−1.25
84	+13.77	+9.52	+7.48	+5.77	+4.42	+5.25	+4.75	+3.75
16	−13.18	−8.77	−6.23	−5.02	−1.68	−4.18	−4.11	−3.96
2	−26.72	−15.25	−8.96	−7.75	−4.42	−5.25	−5.00	−4.50
Cor.	1.134	0.688	0.463	0.356	0.197	0.278	0.234	0.213
% var.	128.28	47.30	21.45	12.67	3.89	7.71	5.47	4.54

The latest standardizations were the best. The earlier ones differ by showing no additional family effects below ages 7 or 8. They do show large family effects at those ages, of course, but still, in almost all data sets the family is more influential before it has to compete with schools and peers. Even taking the sets at face value, the most recent data shows the expected effects. Given the consistency of the latest and intermediate results, with family effects low at the top and high at the bottom beginning at age 12.5, it seems questionable to take the reverse results of the earliest standardizations too seriously.

Above there is another table (Table AIII2, which also appears in the text), in which the recent Raven's results are included. They are derived from Table AV1 with certain percentiles averaged to correspond to the Wechsler percentiles.

References

Ackerman, P. L. (1996). A theory of adult intellectual development: Process, personality, interests, and knowledge. *Intelligence* 22: 227–57.

Adam, S., Bonsang, E., Germain, S., and Perelman, S. (2007). Retirement and cognitive reserve: A stochastic frontier approach to survey data. CREPP Working Paper 2007/04.

Bandura, A. (1993). Perceived self-efficacy in cognitive development and functioning. *Educational Psychologist* 28: 117–48.

Barbey, A. K., Colum, R., Paul, E. J., Chau, A., Solomon, J., and Grafman, J. H. (2014). Lesion mapping of social problem solving. *Brain* 137: 2823–33.

Bouchard, Thomas J. (2013). The Wilson effect: The increase in heritability of IQ with age. *Twin Research and Human Genetics* 16: 923–30.

Capron, C., and Duyme, M. (1989). Assessment of the effects of socio-economic status on IQ in a full cross-fostering study. *Nature* 340: 552–4.

Carroll, John B. (1993). *Human Cognitive Abilities: A Survey of Factor-Analytic Studies*. Cambridge University Press.

Cattell, R. B. (1941). Some theoretical issues in adult intelligence testing. *Psychological Bulletin* 38: 592.

Colum, R. (2014). All we need is brain (and technology). *Journal of Intelligence* 2: 26–8.

Coyle, T. R. and Pillow, D. R. (2008). SAT and ACT predict college GPA after removing "g". *Intelligence* 36: 719–29.

Das, J. P. (2002). A better look at intelligence. *Current Directions in Psychology* 11: 28–32.

Das, J. P., Naglieri, J. A., and Kirby, J. R. (1994). *Assessment of Cognitive Processes*. Needham Heights, MA: Allyn & Bacon.

Deary, I. J., Penke, L., and Johnson, W. (2010). The neuroscience of human intelligence differences. *Nature Reviews Neuroscience* 11: 201–11.

Dickens, W. T., and Flynn, J. R. (2001). Heritability estimates versus large environmental effects: The IQ paradox resolved. *Psychological Review* 108: 346–69.

Duckworth, A. L., and Seligman, M. E. P. (2005). Self-discipline outdoes IQ in predicting academic performance of adolescents. *Psychological Science* 16: 939–44.

Duyme, M. (1981). *Les Enfants abandonnés. Rôle des familles adoptives et des assistantes maternelles*. Paris: CNRS.

Flanagan, D. P. (2014). Cross-battery assessment: A pattern of strengths and weaknesses approach to SLD identification (10/15/2014). Available at www.nyasp.biz/conf_2014_files/Flanagan%20-%20Nov%205.pdf (accessed December 31, 2015).

Flanagan, D. P., Ortiz, S. O., and Alfonso, V. C. (2013). *Essentials of Cross-Battery Assessment* (3rd edn.). New York: Wiley.

Flynn, J. R. (1984). The mean IQ of Americans: Massive gains 1932 to 1978. *Psychological Bulletin* 95: 29–51.

(1987). Massive IQ gains in 14 nations: What IQ tests really measure. *Psychological Bulletin* 101: 171–91.

(2000). IQ gains and fluid g. *American Psychologist* 55: 534.

(2007). *What Is Intelligence? Beyond the Flynn Effect*. New York: Cambridge University Press.

(2008). *Where Have all the Liberals Gone? Race, Class and Ideals in America*. Cambridge University Press.

(2009). Howard Gardner and the use of words. In B. Shearer (ed.), *MI at 25: Assessing the Impact and Future of Multiple Intelligences for Teaching and Learning* (pp. 38–44). New York: Teachers College Press.

(2012a). *Are We Getting Smarter? Rising IQ in the Twenty-First Century*. New York: Cambridge University Press.

(2012b). *Beyond Patriotism: From Truman to Obama*. Exeter: Imprint Academic.

(2013). *Intelligence and Human Progress: The Story of What Was Hidden in Our Genes*. London: Elsevier.

(2015). *Senza alibi: Il cambiamento climatico – impedire la catastrophe* [No place to hide: Spend an evening to learn about climate change]. Turin: Bollati Boringhieri.

Flynn, J. R., te Nijenjhuis, J., and Metzen, D. (2014). The *g* beyond Spearman's *g*: Flynn's paradoxes resolved using four exploratory meta-analyses. *Intelligence* 44: 1–10.

Foulds, G. A., and Raven, J. C. (1948). Normal changes in the mental abilities of adults as age advances. *Journal of Mental Science* 94: 133–42.

Fox, M. C., and Mitchum, A. L. (2013). A knowledge-based theory of rising scores on "culture-free" tests. *Journal of Experimental Psychology: General* 142: 979–1000.

(2014). Confirming the cognition of rising scores: Fox and Mitchum (2013) predicts violations of measurement invariance in series completion between age-matched cohorts. *PLoS One* 9(5): e95780.

Gardner, H. (1983). *Frames of Mind: The Theory of Multiple Intelligences*. New York: Basic Books.

(1993). *Multiple Intelligences: The Theory in Practice, a Reader*. New York: Basic Books.

(1999). *Intelligence Reframed: Multiple Intelligences for the 21st Century*. New York: Basic Books.

(2009). Reflections on my works and those of my commentators. In B. Shearer (ed.), *MI at 25: Assessing the Impact and Future of Multiple Intelligences for Teaching and Learning* (pp. 83–99). New York: Teachers College Press.

Haworth, C. M. A., Wright, M. J., Luciano, M., Martin, N. G., de Geus, E. J. C., van Beijsterveldt, C. E M., Bartels, M., Posthuma, D., Boomsma, D. I., Davis, O. S. P., Kovas, Y., Corley, R. P., DeFries, J. C., Hewitt, J. K., Olson, R. K., Rhea. S.-A., Wadsworth, S. J., Iacono, W. G., McGue, M.,

Thompson, L. A., Hart, S. A., Petrill, S. A., Lubinski, D., and Plomin, R. (2010). The heritability of general cognitive ability increases linearly from childhood to young adulthood. *Molecular Psychiatry* 15: 1112–20.

Heckman, J. J., and Rubenstein, Y. (2001). The importance of non-cognitive skills: Lessons from the GED testing program. *The American Economic Review* 91: 145–9.

Heckman, J. J., Stixrud, J., and Urzua, S. (2006). The Effects of Cognitive and Noncognitive Abilities on Labor Market Outcomes and Social Behavior. NBER Working Paper No. 12006.

Herrnstein, R. J., and Murray, C. (1994). *The Bell Curve: Intelligence and Class in American Life*. New York: Free Press.

Horn, J. L. (1965). *Fluid and crystallized intelligence: A factor analytic study of the structure among primary mental abilities*. Ph.D. thesis. University of Illinois.

Human Brain Project, The (2014). *The Vital Role of Neuroscience in the Human Brain Project* (9 July). Available at www.humanbrainproject.eu/documents/10180/17646/HBP-Statement.090614.pdf (accessed December 31, 2015).

Jensen, A. R. (1970). The heritability of intelligence. *Science and Engineering* 33: 40–3.

(1980). *Bias in Mental Testing*. London: Methuen.

(1998). *The g Factor: The Science of Mental Ability*. Westport, CT: Praeger.

Kelly, R., and Caplan, J. (1993). How Bell Labs creates star performers. *Harvard Business Review* (July–August): 128–39.

Kendler, K. S., Turkheimer, E., Ohlsson, H., Sundquist, J., and Sundquist, K. (2015). The family environment and the malleability of intelligence: A Swedish national home-reared and adopted-away co-sibling control study. *Proceedings of the National Academy of Sciences* 112: 4612–17.

Khaleefa, O., Sulman, A., and Lynn, R. (2009). The increase of intelligence in the Sudan, 1987–2007. *Personality and Individual Differences* 45: 412–13.

McGrew, K. S. (2005). The Cattell-Horn-Carroll theory of cognitive abilities: Past, present, and future. In D. P. Flanagan, J. L. Genshaft, and P. L. Harrison (eds.), *Contemporary Intellectual Assessment: Theories, Tests, and Issues* (pp. 136–82). New York: Guilford.

McGue, M., Bouchard, T. J. Jr., Iacono, W. G., and Lykken, D. T. (1993). Behavioral genetics of cognitive ability: A lifespan perspective. In R. Plomin, and G. E. McClearn (eds.), *Nature, Nurture, and Psychology* (pp. 59–76). Washington, DC: American Psychological Association.

Meisenberg, G. (2014). What are the causes of cognitive evolution? A critique and extension of psychogenetic theory. *Mankind Quarterly* 54: 326–8.

Nisbett, R. E. (2009). *Intelligence and How to Get It: Why Schools and Cultures Count.* New York: Norton.

    (2015). *Mindware: Tools for Smart Thinking.* New York: Farrar, Straus and Giroux.

Oesterdiekhoff, G. W. (2012). Was pre-modern man a child? The quintessence of the psychometric and developmental approaches. *Intelligence* 40: 470–8.

Pinker, S. (2002). *The Blank Slate: The Modern Denial of Human Nature.* London: Penguin.

Raven, J. (2000). The Raven's Progressive Matrices: Change and stability over culture and time. *Cognitive Psychology* 41: 1–48.

Raven, J., Raven, J. C., and Court, J. H. (2003, updated 2004). *Manual for Raven's Progressive Matrices and Vocabulary Scales.* San Antonio, TX: Harcourt.

Raven, J., Rust, J., and Squire, A. (2008a). *Manual: Coloured Progressive Matrices and Crichton Vocabulary Scales.* London: Pearson.

    (2008b). *Raven's Standard Progressive Matrices (SPM) and Raven's Standard Progressive Matrices Plus (SPM Plus).* London: Pearson.

Raven, J. C. (1941). Standardization of progressive matrices. *British Journal of Medical Psychology* 19: 137–50.

References

Raven, J. C., Court, J. H., and Raven, J. (1976). *Manual for Raven's Progressive Matrices and Vocabulary Scales*. London: Lewis. (1986). *Manual for Raven's Progressive Matrices and Vocabulary Scales*. London: H. K. Lewis.

Ritchie, S. J., Bates, T. C., and Deary, I. J. (2015). Is education associated with improvements in general cognitive ability, or in specific skills? *Developmental Psychology* 51: 573–82.

Roid, G. H. (2003). *Stanford-Binet Intelligence Scales: Fifth Edition*. Itasca, IL: Riverside.

Santarnecchi, E., Polizzotto, N. R., Godone, M., Giovannelli, F., Feurra, M., Matzen, L., Rossi, A., and Rossi, S. (2013). Frequency-dependent enhancement of fluid intelligence induced by transcranial oscillatory potentials. *Current Biology* 23: 1449–53.

Schiff, M., Duyme, M., Stewart, J., Tomkiewicz, S., and Feingold, J. (1978). Intellectual status of working-class children adopted early in upper-class families. *Science* 2000, 1503–4.

Schneider, W. J., and McGrew, K. S. (2012). The Cattell-Horn-Carroll model of intelligence. In D. Flanagan and P. Harrison (eds.), *Contemporary Intellectual Assessment: Theories, Tests, and Issues* (3rd edn., pp. 99–144). New York: Guilford.

Staff, R. T., Hogan, M. F., and Whalley, L. J. (2014). Ageing trajectories of fluid intelligence in late life: The influence of age, practice and childhood IQ on Raven's Progressive Matrices. *Intelligence* 47: 194–201.

Sternberg, R. J. (1988). *The Triarchic Mind: A New Theory of Human Intelligence*. New York: Penguin. (1997). *Successful Intelligence: How Practical and Creative Intelligence Determine Success in Life*. New York: Plume. (2006). The Rainbow Project: Enhancing the SAT through assessments of analytic, practical, and creative skills. *Intelligence* 34: 321–50.

Sternberg, R. J., Forsythe, G .B., Hedlund, J., Horvath, J. A., Wagner, R. K., Williams, W. M., Snook, S. A., and Grigorenko, E. L. (2000). *Practical intelligence in everyday life*. New York: Cambridge University Press.

Thorndike, R. L., Hagen, E. P., and Sattler, J. M. (1986). *Stanford-Binet Intelligence Scale: Fourth Edition.* Chicago: Riverside.

US National Institute of Health (2014). *BRAIN 2025: A scientific vision.* BRAIN Working Group report to the Advsiory Committee to the Director, NIH, June 5, 2014. Washington, DC. Available at www.braininitiative.nih.gov/2025 (accessed December 31, 2015).

van der Maas, H. L. J., Dolan, C. V., Grasman, R. P. P. P., Wicherts, J. M., Huizenga, H. M., and Raijmakers, M. E. J. (2006). A dynamical model of general intelligence: The positive manifold of intelligence by mutualism. *Psychological Review* 113: 842–61.

Wechsler, D. (1949). *Wechsler Intelligence Scale for Children: Manual.* New York: The Psychological Corporation.

(1955). *Wechsler Adult Intelligence Scale: Manual.* New York: The Psychological Corporation.

(1974). *Wechsler Intelligence Scale for Children – Revised.* New York: The Psychological Corporation.

(1981). *Wechsler Adult Intelligence Scale – Revised.* New York: The Psychological Corporation.

(1989). *Wechsler Preschool and Primary Scale of Intelligence – Revised.* San Antonio, TX: The Psychological Corporation.

(1992). *Wechsler Intelligence Scale for Children – Third Edition: Manual* (Australian Adaptation). San Antonio, TX: The Psychological Corporation.

(1997). *Wechsler Adult Intelligence Scale – Third Edition: Manual.* San Antonio, TX: Pearson.

(2002). *Wechsler Preschool and Primary Scale of Intelligence – Third Edition: Manual.* San Antonio, TX: Pearson.

(2003). *Wechsler Intelligence Scale for Children – Fourth Edition: Manual.* San Antonio, TX: The Psychological Corporation.

(2008). *Wechsler Adult Intelligence Scale – Fourth Edition: Manual.* San Antonio, TX: Pearson.

Wood, R. E., and Bandura, A. (1989). Impact of conceptions of ability on self-regulatory mechanisms and complex decision-making. *Journal of Personality and Social Psychology* 56: 407–15.

Woodley, M. A. (2012a). A life history model of the Lynn-Flynn effect. *Personality and Individual Differences* 53: 152–6.

(2012b). The social and scientific temporal correlates of genotypic intelligence and the Flynn effect. *Intelligence* 40: 189–204.

Woodley, M. A., Figueredo, A. J., Ross, K. C., and Brown, S. D. (2013). Four successful tests of the cognitive differentiation-integration effort hypothesis. *Intelligence* 41: 832–42.

Name index

Subject index